PRIVATE SECURITY

Volume I

Bruce George MP
Mark Button

Perpetuity Press

Published by
Perpetuity Press Limited

PO Box 376,
Leicester, LE2 1UP, UK
Telephone: +44 (0) 116 221 7778
Fax: +44 (0) 116 221 7171

214. N. Houston
Comanche, Texas 76442, USA
Telephone: 915 356 7048
Fax: 915 356 3093

Email: info@perpetuitypress.co.uk
Website: www.perpetuitypress.com

The views expressed in this book are those of the authors and do not necessarily reflect those of Perpetuity Press Limited.

British Library Cataloguing in Publication Data.
A catalogue record for this book is available from the British Library.

Private Security
Bruce George and Mark Button

ISBN 1-899287-70-1

Perpetuity Press

This book is dedicated to our parents:
the late Phyllis and Edgar George,
Andrew and Sally Button

About the authors

Bruce George

Bruce George is the Member of Parliament for Walsall South. He has been campaigning for regulation of the private security industry since 1977 and has introduced five private members bills and spoken extensively on this issue in parliament. He has written numerous papers on private security issues and is a frequent contributor to international conferences on this subject. He is also currently Chairman of the House of Commons Defence Committee.

Mark Button

Mark Button is a Senior Lecturer at the Institute of Criminal Justice Studies, University of Portsmouth. He is course leader for the MSc in Criminal Justice Studies and the Certificate of Higher Education in Risk and Security Management. He has written extensively on private security issues and has spoken at numerous international conferences on the subject. Prior to becoming a lecturer, he was a Research Assistant to Bruce George.

Contents

LIST OF FIGURES AND TABLES

Figures

Tables

Foreword

Rt Hon Jack Straw MP

I am delighted to have been invited to contribute the foreword to this book on the British security industry. The authors, Bruce George and Mark Button, are to be congratulated on producing such a comprehensive contribution to understanding the industry.

It is easy to overlook the long history of the private security industry. Today, it is a strong, modern and expanding industry, which is performing an increasingly prominent and important role in our society. There are, and will continue to be, opportunities for greater involvement too. The Government's commitment to a partnership approach to tackling crime and disorder has paved the way for the private security industry to play a wider role in initiatives to reduce crime and improve community safety.

We are fortunate in having the security industry that we have. Its role has become more obvious to every one of us, through the visible presence of security staff in both public and private places, and through the growing use of security aids such as CCTV and alarm systems.

But the progress of the industry has been marked not only by an increase in scale but also by growing professionalism, diversification, and innovation. The industry is to be greatly applauded for the efforts that it has made to increase its professionalism, raise its standards and improve its standing with the public and with fellow professionals. We are anxious to build on this progress.

To achieve a greater level of partnership, and to make it work effectively, we need to establish pre-conditions that will ensure mutual confidence between Government, industry, the police, local authorities and the public. Regulation of the industry will help provide a stronger basis for the future. We need a well-trained industry with high standards across the board; we need regulation to remove the criminal elements from the industry who tarnish its image; and we need an industry that can fully enjoy the confidence of the public. These are the key principles underpinning our proposals for regulation of the industry, which the authors discuss towards the end of their book.

We published our proposals in a White Paper last year. They provide for the establishment of a new Authority whose responsibilities will be to protect the public by ensuring basic standards of probity within the industry; and maintaining and improving standards. We have made clear our intentions to bring forward legislation as soon as parliamentary time is available. This book will greatly inform the debate.

Acknowledgements

There have been so many people involved in helping us write this book that a list could almost take up a chapter in its own right; and it is not possible to thank everyone who has helped us. There are some, however, who deserve special mention. First, we must thank some of the many individuals who allowed us to interview them and provided us with much documentation to read — in particular, Dennis Bellamy, John Casson, Raymond Clarke, Mervyn David, Ian Drury, Eric Ellen, David Fletcher, Peter French, Dr Martin Gill, David Hinge, David Holt, Lord Imbert, Ian Jack, Russell Jenkins, Professor Les Johnston, Peter Jones, Stewart Kidd, The Rt Hon Alun Michael MP, Ann Perkins, Professor Robert Reiner, Ian Sanderson, John Smith (Prudential), Sir John Smith (Former Metropolitan Police Service Deputy Commissioner), David Sowter, Sir John Stevens, Dennis Stopford, The Rt Hon Jack Straw MP and Mick Upton, amongst many others. We must also thank the following organisations and companies for the help they provided to us: ABI, ACPO, ASC, ASIS, BRC, BSIA, Control Risks, EPIC, GMB, ISM, NACOSS, Pinkerton, Rentokill, the Scarman Centre, Securicor, SITO and the West Midlands Police, amongst others. We must also thank Barry Loveday, Mark Stenhouse, Nick Ryan and David Dickinson, for their comments on earlier drafts. The House of Commons Library also proved invaluable in securing books and information — particular thanks must go to Mary Baber and Andrew Parker — as did many local libraries and archives, especially Walsall's. During the writing of this book there were also a number of research assistants who contributed to the overall research, and we would like to thank Maria Burley, Simon Cooper, Ian Geary, Geoff Smith and Jonathan Whitehouse. Special thanks are also due to Group 4, and in particular to David Dickinson and Jim Harrower, for their extensive co-operation. The authors would also like to thank Group 4 for its permission in using the photograph of one of their control rooms on the front cover. Finally we must also thank Karen Gill for her patience throughout this project.

Glossary of Acronyms and Initials

ABI	Association of British Investigators
ABIn	Association of British Insurers
ABIS	Association of Burglary Insurance Surveyors
ACPO	Association of Chief Police Officers
AISC	Alarm Inspectorate and Security Council
AMA	Association of Metropolitan Authorities
APPSS	Association of Police and Public Security Suppliers
ASC	Association of Security Consultants
ASIS	American Society for Industrial Security
ATM	Automatic Teller Machine
AUCSO	Association of University Chief Security Officers
AUSSS	Assistant Under-Secretary Security and Support
BCS	British Crime Survey
BEDA	British Entertainment and Discotheque Association
BIDT	British Institute of Dog Trainers
BLMA	British Lock Manufacturers Association
BMA	British Medical Association
BRC	British Retail Consortium
BS	British Standard
BSIA	British Security Industry Association
CAA	Civil Aviation Authority
CBAEW	Certificated Bailiffs Association of England and Wales
CCTV	Closed Circuit Television
CCTVMDA	Closed Circuit Television Manufacturers and Distributors Association
CEN	Commission for European Norms
CENELEC	Commission for European Norms Electrical
CIT	Cash-in-Transit
CJA	Criminal Justice Act
CJPOA	Criminal Justice and Public Order Act
COESS	Confederation of European Security Services
CPP	Certified Protection Professional
CRISA	Car Radio Industry Specialist Association
DCMF	Designed, Constructed, Managed and Financed
DERA	Defence Evaluation and Research Agency
DETR	Department of the Environment Transport and the Regions
DIY	Do-It-Yourself
DSMA	Door and Shutter Manufacturers Association
DVST	Digital Video Storage and Transmission
EAS	Electronic Article Surveillance

EASEM	European Association of Security Equipment Manufacturers
EASMA	Electronic Article Surveillance Manufacturers Association
ECA	Electrical Contractors Association
ECAC	European Civil Aviation Conference
ECSA	Express Carriers Security Association
EFSG	European Fire and Security Group
ELF	European Locksmiths Federation
EN	European Norm
EP	Evaluation Panel
EPIC	Ex-Police in Industry and Commerce
ESF	European Security Forum
ESTA	European Security Transport Association
EURALARM	European Committee of Alarm Manufacturers and Installers
EUROSAFE	European Committee of Safe Manufacturers Associations
EVSA	European Vehicle Security Association
FAA	Federal Aviation Administration
GMAF	Greater Manchester Alarms Federation
GSA	Group Security Adviser
GSC	Guild of Security Controllers
GSM	Group Security Manager
HAC	House of Commons Home Affairs Committee
HCDC	House of Commons Defence Committee
HMIP	Her Majesty's Inspectorate of Prisons
HO	Home Office
IATA	International Air Transport Association
IBA	International Bodyguards Association
ICAO	International Civil Aviation Organisation
IDC	Immigration Detention Centre
IHSM	Institute of Hotel Security Management
IIS	International Institute of Security
IKD	International Federation of Associations of Detectives
IND	Immigration and Nationality Department
IPI	Institute of Professional Investigators
IPSA	International Professional Security Association
ISI	Inspectorate of the Security Industry
ISM	Institute of Security Management
ISMA	International Security Management Association
JCWI	Joint Council for the Welfare of Immigrants
JSIC	Joint Security Industry Council
LGDIS	Liaison Group for Defence Industrial Security
LISS	Ligue Internationale de Sociétés de Surveillance

LPC	Loss Prevention Council
LPCB	Loss Prevention Certification Board
LPS	Loss Prevention Standard
MDP	Ministry of Defence Police
MESF	Mobile Electronics and Safety Federation
MGS	Ministry of Defence Guard Service
MLA	Master Locksmiths Association
MLSE	Military Local Service Engagements
MoD	Ministry of Defence
MP	Member of Parliament
MPGS	Military Provost Guard Service
NACCB	National Accreditation Council for Certification Bodies
NACOSS	National Approval Council for Security Systems
NAID	National Association for Information Destruction
NAHS	National Association for Healthcare Security
NALDS	National Association of Licensed Door Supervisors
NAO	National Audit Office
NARDS	National Association of Registered Door Supervisors
NARDSSP	National Association of Registered Door Staff and Security Personnel
NASDU	National Association of Security Dog Users
NASP	National Aviation Security Programme
NASS	National Association of Security Services
NCIS	National Criminal Intelligence Service
NFTA	National Fencing Training Authority
NHS	National Health Service
NSCIA	National Supervisory Council for Intruder Alarms
NVQ	National Vocational Qualification
PAC	Public Accounts Committee
PCO	Prison Custody Officer
PF/PSI	Police Foundation/Policy Studies Institute
PIRA	Provisional Irish Republican Army
PSDB	Police Scientific Development Branch
PSIA	Private Security Industry Authority
PUS	Permanent Under-Secretary
RCS	Retail Crime Survey
RM	Royal Marines
RN	Royal Navy
RSMF	Risk and Security Management Forum
RSPCA	Royal Society for the Prevention of Cruelty to Animals
SAFE	Security and Facilities Executive
SII	Security Institute of Ireland
SILB	Security Industry Lead Body
SITO	Security Industry Training Organisation
SSA	Security Services Association

SSAIB	Security Services and Alarms Inspection Board
SSI	Security Services Inspectorate
TRANSEC	Transport Security Division
UKAEC	United Kingdom Atomic Energy Authority Constabulary
UKAS	United Kingdom Accreditation Services
USDAW	Union of Shop, Distributive and Allied Workers
VSIB	Vehicle Security Installation Board

Part 1. Introduction

Chapter 1

Introduction

If, at the end of the 1950s, a futurologist had predicted to a policeman that by the turn of the century there would be more private security officers than police officers, that they would be routinely patrolling public areas; that security firms would be managing and owning half a dozen or more prisons as well as transporting prisoners; that most town centres and some residential areas would be ordinarily under close circuit television (CCTV) surveillance; and that police forces would be increasingly charging for some of their services and even be sponsored by business, the policeman would probably have thought he was mad. All these predictions, however, have come true and they illustrate the fundamental change that has taken place in the structure of policing. At the end of the Second World War, policing and its related functions were the virtual monopolies of the state, and the private security industry hardly existed (although, as chapter 3 will illustrate, there were many comparable security roles). This has changed and Weber's (1964) notion of the state as possessing the monopoly of coercion in a given area is no longer as convincing — if it ever was.

The involvement of private organisations in policing is not new, however, and in many cases such involvement pre-dates the formation of the modern police in 1829. What is distinct is their phenomenal growth over the last 25 years and the transformation of some of their activities into the private security industry. Such has been the growth of private security in some countries that there are now more private security officers than police officers, leading Cunningham, Strauchs and Van Meter (1990) to describe private security in the USA as the nation's 'primary protection resource'. Moreover, when the activities of the police are set against those of the private security industry in the UK, there are very few of the former that are not now undertaken by the private sector (Jones and Newburn, 1998).

However, despite the growth in the size and role of the private security industry, there has been little research in the UK in comparison to that undertaken into public policing bodies (Reiner, 2000). First, there is a community of academics dedicated to research into the police in numerous university police, criminology, law, and politics departments. There is a range of first degree and master's level qualifications in subjects related to the police, policing, and criminal justice. There is also a burgeoning body of literature in both books and academic journals. By comparison, there are only a handful of academics interested in private policing, a few universities offering degrees in security management and a limited — although growing — body of academic literature.

In the UK over the last 30 years, there have only been six major books that have attempted to undertake a broad and deep look at the private security industry. Clayton (1967) offered the first significant review, although focusing mainly on the cash-in-transit (CIT) sector. The next major study was undertaken by Draper (1978), who took a much wider view of the private security industry. *Policing for Profit*, written by Nigel South in 1988, pursued a similarly broad view to Draper's, but like her did not review in depth some important sectors of the industry such as: door supervisors, security consultants, and the manufacturers, distributors and installers of security equipment. Johnston (1992a) reviewed private security from the broader perspective of non-police policing, assessing those public bodies other than the police engaged in policing, as well as other private phenomena. More recently, McManus (1995) has published research into public patrols by private security officers. Finally, Jones and Newburn (1998) have published the results of their extensive research into the private security industry. Again, however, the research focuses largely upon private security officers engaged in public policing roles.

In some other countries there has been more extensive research into the private security industry. In the USA over the last 30 years there have been a number of significant government-funded reports. The 1971 five-volume Rand report is probably the most extensive study of private security that has ever been undertaken (Kakalik and Wildhorn, 1971a, b, c, d, e). It was followed less than five years later by another significant government-funded research project, conducted by the National Advisory Committee on Criminal Justice Standards and Goals (1976). These have been succeeded by the Hallcrest Reports 1 and 2 (Cunningham and Taylor, 1985; Cunningham et al, 1990). The USA is also home to a large academic community dedicated to research into private security, which offers numerous subdegree, first degree and master's level qualifications. Therefore it is not surprising to find academic books, such as Fischer and Green's (1992) *Introduction to Security*, which review private security in the USA and which are aimed at those studying for the many qualifications available there.

However, perhaps the most academically significant research on private security was undertaken across the border in Canada, at the University of Toronto's Centre for Criminology, through the work of Philip C. Stenning and Clifford D. Shearing (see for instance, Shearing and Stenning, 1982; 1987). Unfortunately, much of their work was undertaken in the late 1970s and early 1980s. In Australia, the growth of private security has also recently stimulated a burst of research activity in this field, although focused more on private prisons; some of the more significant works include Biles and Vernon (1994), Moyle (1994), Sarre (1994) and Prenzler and Sarre (1999).

There are also a number of books dedicated to a particular sphere of security, although aimed more at practitioners than academics: for instance, in the field of aviation security Moore's (1991) *Airport, Aircraft and Airline Security* and Dorey's (1983) *Aviation Security*. In the area of retail security, Beck and Willis's (1995) *Crime and Security: Managing the Risk to Safe Shopping* and Gill's *Crime at Work I* (1994) and *Crime at Work II* (1998). In the USA there are many other titles dedicated to areas such as campus, hotel and hospital security, to name but a few.

Therefore, given the increasing importance of private security in the UK and the lack of detailed studies, we decided to go about researching it, with the aim of publishing a book that fills this gap. We also wanted to undertake a detailed analysis of the *wider* private security industry, and not just to focus on the popular activities such as security officers and private investigators, as some earlier studies have. In undertaking such a study, we felt it was also important to examine how private security is used in commerce and the public sector. For instance, what differences are there in the use of private security in the aviation, retail and other sectors? Another motive for publication has been frustration at the lack of enthusiasm for, indeed indifference towards, the regulation of the private security industry. Therefore this book is the first — that we are aware of — to combine a sectoral study of the industry with a description of how private security is actually used. As statutory regulation approaches, it is also hoped this book will contribute to the shape of the long awaited legislation.

We commence Part 1 by considering the nature of 'private security' through a brief analysis of its origins, combined with an examination of its remarkable growth in the latter half of the twentieth century. In doing so, we also attempt to develop a definition of what the private security industry is, and to examine some of the many reasons for its remarkable growth.

In Part 2 we then provide a detailed analysis of the different sectors that we argue constitute the private security industry, illustrating its increasingly fragmented nature. We start by considering the many organisations that are involved in the industry, then move on to consider its distinct sectors: manned security services, private sector detention services, security storage and shredding services, professional security services, security products, and the margins of the industry.

In Part 3 we explore the integrated nature of private security through different case studies from both commerce and the public sector. These illustrate the fact that although the industry may be fragmenting in terms of organisations and sectors, when security decision-makers decide upon a security strategy to address a risk, they generally consider applying it in an integrated manner, using a wide range of products and services from the different sectors of the industry. To undertake this analysis, we begin Part 3 by considering the security decision-maker, before examining security in the retail, aviation and entertainment sectors and in the Ministry of Defence (MoD).

In Part 4, the final section of the book, we focus on the way in which private security has been considered as a policy issue in the UK and Europe and, most notably, on the issue of regulation, which is likely to become increasingly important for the industry in the future.

In this volume the focus will largely be on the UK, although there will be references to international examples where necessary. In Volume 2 international perspectives on the private security industry will be considered, including the assessment (amongst many issues) of how the characteristics and regulatory structure of the private security industry vary in different parts of the world.

Chapter 2

Private Security Defined

For most people, the image of private security is one of armoured cars outside banks or uniformed guards patrolling shopping centres. Private security, however, encompasses more than this. There are also, for instance, alarm installers, locksmiths, private investigators and manufacturers of security equipment. A good analogy for the private security industry is the familiar iceberg, with the high-profile uniformed sector the tip above the surface and the rest of the industry — with its lower profile but immense size — below. There are also many different perceptions of what constitutes private security, which makes defining it very difficult. Furthermore, there are many grey areas at the margins, which may or may not be regarded by different individuals and organisations as within the scope of private security. Hence there are many obstacles to creating an acceptable definition. This chapter will consider some of the existing definitions, illustrate some of their inadequacies and then move on to attempt to develop a definition of its own.

The definition of the word 'security' is problematic, since it is commonly used in a number of different ways. For instance, it is often used in international politics in the context of relations between countries, and also in that of the protection of the nation state. For instance, MI5 and MI6 (as they were previously known) were described as 'the security services'. The British government often refuses to release information or takes action on the grounds of 'national security'. The word 'security' is also used to mean collateral ('security against a loan'), or welfare payments ('social security'). In French, our separate English notions of security and safety are combined in the single noun *sécurité*, which covers them both (Gill, 1996). The wide range of meanings that the word 'security' can carry is illustrated by a dictionary definition, such as that in the *New Shorter Oxford English Dictionary*:

> The condition of being protected from or not being exposed to danger ... The condition of being protected from espionage, attack, or theft. Also, the condition of being kept in safe custody ... The provision or exercise of measures to ensure such safety. Also, a government department or other organisation responsible for ensuring security ... Freedom from care, anxiety, or apprehension; a feeling of safety or freedom from danger ... Formerly also, over confidence, carelessness ... Freedom from doubt; confidence, assurance ... The quality of being securely fixed or attached, stability.

Thus the word is used in a wide range of contexts beyond the activities to be considered in this book, which deals only with those generally taken to come under the term 'private security industry'. This wider use further compounds the difficulties of defining the term, though there have been a number of attempts. They can be broadly divided into those that list the activities and sectors believed to be part of the industry, and those that define it in terms of relevant functions. Both strategies are fraught with problems and no one definition, in our view, has yet proved adequate. It might also be useful at this point to distinguish between the 'private security industry' and 'private security': the former encompasses all those individuals and organisations that produce and provide 'private security' products and services; the latter comprises the products and services themselves.

'List-based' definitions

The easiest route to an attempted definition, although not without its pitfalls, is to list those activities that are believed to be part of the industry (George and Watson, 1992; Home Office, 1979; Kakalik and Wildhorn, 1971a; Schiffi, 1989; Trofymowych, 1993). The differing scope of definitions is illustrated by the following two. The first is that of Trofymowych (1993:3) who divides the industry between manned security and security hardware, defining the former as:

> the provision of personnel who perform policing functions such as guards, patrol persons, floor detectives, investigators, escorts, couriers, alarm respondents, auditors and security consultants.

In the second definition Kakalik and Wildhorn (1971a:3) argue that:

> Private police and private security forces and security personnel are used generically in this report to include all types of private organisations and individuals providing all types of security related services, including investigation, guard, patrol, lie detection, alarm, and armoured transportation.

Such definitions offer no rationale as to why particular sectors should be included, and are effectively saying that these are the activities or sectors of the private security industry, and that therefore they define that industry. This is akin to saying 'this is a car because I say it is', without offering the characteristics that define such vehicles. The task is to find a definition which can be applied to assess whether something is or is not part of private security. This should be possible, for if it was not, as Manunta (1999) has argued, it would not be possible to distinguish truth from falsehood.

'Function-based' definitions

It is not surprising that there exist 'list-based' definitions of the private security industry, given its vast scope and given the difficulties of constructing a 'function-based' definition that is neither so wide as to include industries and professions that most would not consider part of the industry, nor so narrow as to exclude sectors that would generally be considered as within its scope. The challenge is therefore to find a definition by which all activities and sectors can be judged, in deciding whether or not they constitute part of the private security industry. As Shearing and Stenning (1981:195) write:

> What is perhaps more useful is to identify the unifying theme which makes all these services 'security services'. That theme, we believe, is protection against depredation, and in particular the protection of information, persons and property.

Jordans and Son (1992: VII) develop such themes further, defining the industry as:

> concerned with the protection of physical property, assets and individuals from theft or violence. In its simplest analysis this protection can be categorised into three sections, namely by the application of (a) physical/mechanical devices, (b) electrical/electronic apparatus, (c) manned services, or, of course, various combinations of the three.

More recently the Home Office (1995a:81), in its submission to the Home Affairs Committee (HAC) inquiry into the private security industry, defined the latter in the following terms:

companies which employ staff to protect property and individuals, or that install systems to do so. There are distinctions to be drawn between the private security industry and public security agencies (such as the police and prison service) in terms of accountability and legal authority. But there are, increasingly, common areas of activity and contracting out has meant that some previously public sector tasks are now performed by private companies.

Group 4 (1992: 13) has taken a very different approach to defining professional security — placing it as part of risk management, namely:

> that part of risk management which seeks to reduce the chance by preventing, detecting, or dealing with fire, crime, accident and waste (often called 'loss prevention').

Jordans and Son illustrate the problems of creating a definition which is too narrow. By restricting the industry to the '... protection ... from theft or violence' they exclude the large sections of it that are involved in protection from other crimes, such as fraud and vandalism. The definition also focuses on protection from crime, though many security officers are employed to perform other functions unrelated to crime, such as crowd control, public order functions and fire watch duties. Most notably, the definition would exclude private investigators, which many would find odd.

The Shearing and Stenning, Home Office, and Group 4 definitions, on the other hand, can be criticised as too wide. For instance, a definition based on 'protection' might include a pest control company, for such a service would provide greater security to an individual by protecting against pests such as wasps, rats, etc. Perhaps such a definition could also be stretched to include a decorator who paints windows with protective varnish (since a rotten window frame could be construed as a threat to the building's security). The Shearing and Stenning definition would also include the public security sector and the Home Office definition would exclude the many entities (sole traders, partnerships, etc) which are not companies but which operate within the industry. Similarly, the Group 4 definition, by including the threats of fire and accident within the scope of risk management, also extends the scope of private security to functions such as those carried out by safety and fire officers, and to products such as fire alarms, that most would not regard as part of the industry. These examples illustrate the minefield of attempting to define the private security industry through such notions as 'protection'.

Thus, defining the industry presents immense problems, because there are a wide range of activities and products that may or may not fall within its scope. Given that no definition devised so far successfully defines the industry in terms of a function, the challenge is to create one — if possible — that does.

Towards a definition?

The essence of private security

The first problem is to draw a line dividing where private security ends and other areas begin. Would bailiffs, credit reference agencies, installers of fire alarms and wheel clampers be included, for instance? This list could be extended to several pages, and make equally convincing cases for including them within, or excluding them from, private security. The challenge is to find a collection of words that is wide enough to capture all the relevant sectors, without being so wide as to destroy the precision of the definition.

Some of the definitions illustrated earlier used the notion of 'protection', which was shown to be either too wide or too narrow. Other concepts such as crime prevention, loss prevention, loss reduction, safety, risk reduction and even security have also been used. However, there are drawbacks to all these concepts. 'Crime prevention' restricts private security to crime-related functions, when there are quite clearly security personnel engaged in functions outside crime prevention. 'Loss reduction' and 'loss prevention' seem to restrict private security to the business environment. 'Safety' shifts the focus to a more marginal activity than that which most people in the private security industry are involved in, as does 'risk management'. Using 'security' to define private security (as Shearing (1981) does — he defines private security as 'persons employed in security occupations other than the military and public police, and other public officials with peace officer status') also does little to improve our understanding of what constitutes private security.

Ultimately, most academics and practitioners start with preconceptions of what constitutes private security, such as security officers, private investigators, alarm installers, etc, but not pest control organisations, fire officers, fire alarms, insurance companies. Unless one is prepared to change one's existing preconceptions of what private security is, the use of different concepts will always create a definition of private security that differs from the pre-conceptions of some, and will therefore be unacceptable to them.

Nevertheless, a number of functions can be identified that link most private security services and products. They include crime prevention, order maintenance, loss prevention, and protection — though these are not common or exclusive to all private security products and services. Thus the starting point for an analysis is the extent to which one or more of these functions characterises the product or service in question; and the more functions that do characterise it, the more clearly it can be seen as part of the private security industry.

Figure 1. The essence of private security

Role of product or service
Crime prevention Order maintenance Loss prevention Protection
The more these functions characterise a product or service, the stronger its claim to be part of private security.

If, then, this criterion was applied to pest control companies, it would be clear that only protection would be relevant, with possibly loss prevention in some circumstances. However, if it were applied to a typical private investigator then all four functions would be relevant. We do not think it is possible to draw an absolute line at the point where an activity becomes private security; rather, the more of the above functions that are relevant to that activity, the more clearly it can be viewed as private security. As will shortly be illustrated in the case of the public/private divide, it is more appropriate to view private security in terms of the degree of 'private securityness', although where most activities that we are aware of within the private security industry are concerned, at least three of the functions are relevant (see Table 1).

The private security industry is also divided into a number of distinct sub-sectors according to the service and product provided, to the market served and to the regulatory structure, among other factors. These divisions have been described in many reports and also by some academics (South, 1988; Johnston, 1992a; Jones and Newburn, 1998). For instance, Jones and Newburn distinguish staffed services, security products and investigation services. We, however, wish to differentiate the industry further and, basing our analysis on the different characteristics of these sub-sectors (which will be explored in chapters 5 to 11), have arrived at the following taxonomy (see Table 1). In our view, the Table also illustrates the degrees of 'private securityness'.

Table 1. A taxonomy of the private security industry

Sector	Crime Prevention	Order Maintenance	Loss Prevention	Protection
Manned security services				
Static guarding	*	*	*	*
Cash-in-transit	*		*	*
Door supervisors/stewards	*	*	*	*
Close protection	*	*	*	*
Detention services	*	*		*
Security storage/shredding	*		*	*
Professional security services				
Security consultants	*		*	*
Professional investigators	*	*	*	*
Security Products				
Intruder alarms	*		*	*
CCTV	*	*	*	*
Access control	*	*	*	*
EAS equipment	*		*	*
Detection equipment	*		*	*
Locks	*		*	*
Safes/ vaults/ security storage	*		*	*
Barriers/ shutters	*		*	*
Fences	*		*	*
Security glass/windows	*		*	*
Armoured vehicles	*		*	*

Notes:
1. **Bold** type denotes sector, *italics* a part of a sector. An asterisk * means that the function is relevant to the sector.
2. This is an indicative list — not all services/products are included.
3. Detention services are a special case for inclusion within the bounds of the industry; the rationale for this will be explored later in this chapter.

The public/private dichotomy

The public — private dichotomy has been the subject of much academic debate (Jones and Newburn, 1998). A number of writers have attempted to separate out the factors which distinguish private from public. Pitkin (1984) has identified as indicators of the distinction: the degree of accessibility; the extent to which something (a product or service) affects us all and the extent to which that something is covered by public administration or the collective state. However, as Jones and Newburn show, it is probably more appropriate to talk about the degree of 'privateness' or 'publicness', given the many grey areas that exist, rather than imposing an either/or choice. In doing so, these authors distinguish a number of illustrative characteristics, including the following: the mode of provision (collective or otherwise); sources of funding; the nature of the relationship between provider and user and the employment status of the personnel concerned. Ultimately, there are many ways in which the degree of 'publicness' and 'privateness' can be assessed. For the purposes of this chapter these and other issues will be drawn upon to create the most effective means of distinguishing the 'private' in private security. To further complicate the issue, however, Johnston (2000) has suggested that 'commercial' security might be more appropriate than 'private'. However, as 'commercial' would seem to exclude the public sector, and possibly in-house security departments in the private sector, we have decided to continue with 'private', despite the difficulties which this term also raises.

Thus the question to address in considering the private security industry is where the boundary lies between the private and public sectors. Security officers are increasingly found patrolling shopping centres and streets, functions traditionally associated with the police. Outside the UK, private security has actually replaced police officers when the latter were on strike (Shearing, Stenning and Addario, 1985). Police forces are also competing for work with the private sector in some areas of activity, and there are also many organisations in the public sector which carry out security/policing roles, but which do not belong to one of the Home Office police forces. These issues lead to the question of where to draw the line between the public and private sectors.

Almost all writers and commentators on the industry refer to it as 'the private security industry'. The 'private' in the title would seem to imply that any activity in the public security sector is excluded. This seems reasonable when the public security sector referred to is the police, the armed forces, etc. However, there are many individuals directly employed within the public sector who undertake duties almost identical to those carried out by security officers of private security companies; these include the MoD Guard Service, Metropolitan Police Security Guards, in-house security in government departments and agencies, and the many in-house security officers employed by local authorities. This raises the question of whether these should also be considered as part of the private security industry. After all, if we were to consider an intruder alarm at a military establishment (which is certainly part of the 'public' security apparatus) we would still consider it a product of the private security industry.

This line of argument can also include certain investigators in the public sector, for example, health and safety inspectors or local authority environmental health inspectors, within the bounds of private security, since after all such people undertake investigations similar to those of private investigators. Most writers and commentators on private security would regard them as outside the scope of private security; however, detailed analysis of the roles of different investigators in the private sector would reveal some with comparable roles to certain public sector investigators, such as housing benefit fraud investigators for a local authority, and others with more enforcement-like roles related to the police, such as environmental health inspectors.

There is also a common theme that unites security personnel working in the private sector with those in many public bodies, and distinguishes them from the police and other public law enforcement agencies: they serve a private interest, rather than the public interest served by most police officers, customs officers, health and safety inspectors, etc (although one could also engage in a lengthy debate on what constitutes public as opposed to private interest). Thus a security officer employed by a local authority to guard and patrol a block of flats would generally be serving the residents' interest — a private interest, whereas a policeman who decided to patrol the flats would be generally serving the public interest.

The final distinguishing factor is whether the personnel concerned are vested with any special statutory powers. There are now some personnel working for private security companies that possess such powers — prison custody officers working in private prisons or in private prison escort services, for example. Generally, however, most personnel working in the private security industry do not have any special statutory powers. Thus whether an individual has such powers is a good means of distinguishing whether a person with a security role employed by the public sector could be considered as private security, on the margins, or not part of it at all.

Thus the questions defining whether a security role in the public sector can be considered as 'private security' are first: can a comparable occupation can be found in the private security sector? If not, the role is not part of the private security industry. Second, whose interest does the function serve? If this is a private interest, it is potentially a part of the private security industry. Third and finally, do the personnel have any special statutory powers? If so, they should also not be considered as private security. If the answers to questions two and three do not consistently fit either private security or public security, then the occupation could be described as in the grey area we have called 'the margins' between the private security industry and public security sector. Figure 2 illustrates how to assess whether public sector employees in security roles can be considered as private security.

Figure 2. A test for whether public sector employees can be considered as private security

a. Nature of job
Comparable security role to private security sector occupation?
No   Yes

b. Interest served
Private   Public

c. Special powers
No   Yes

Private security   *The margins*   *Public security*

Thus if (a) a public sector role is comparable to a private sector role, (b) the interest served is private and (c) there are no special statutory powers, then that role can be considered as private security. For example, if one considers the role of an investigator employed by the Post Office to investigate fraud and the counterfeiting of stamps, the following questions need to be asked: first, is there a comparable job in the private sector? Yes, as many private investigators are involved in investigating fraud and counterfeiting. Second, whose interest does the investigator serve? This is not as clear-cut, since preventing fraud in a public service could be seen as serving the public interest by reducing public revenue losses. However, where the police serve all interests, an investigator in the Post Office would ultimately serve the Post Office's interest, and for this reason he/she can be seen as serving a private interest. Finally, does the investigator have any special statutory powers? The answer is no, and therefore an investigator employed by the Post Office could be considered as part of the private security industry. Another example would be security officers employed by the Metropolitan Police at the Palace of Westminster. Their role is comparable to those in the private sector, the interest they serve is private, as they serve the Houses of Parliament, and they have no special statutory powers. Hence they can be described as a form of private security. There are many other services, however, that do not pass this test so clearly, and would fall within the margins between the private and public sectors.

There are also uncertain areas in the private sector. These have been expanded by the creation of the private prison and prison escort sector. There are now employees of private organisations, with roles that were once monopolised by the public sector, who serve the public interest (ie, by keeping in prison those that the state has decided are a risk to society) and who have special powers. However, because these personnel are employed in the private sector by a majority of companies that are essentially private security businesses, they have been considered as part of (and a distinct sector of) the private security industry. Another example of the grey area in the private sector is an organisation such as the Royal Society for the Prevention of Cruelty to Animals (RSPCA), which carries out functions in the private domain that are similar to those of police officers, ie, the investigation of crimes, gathering of evidence and prosecution of individuals. In protecting animals, they serve the public interest by providing a service to the whole of society, although they have no special statutory powers. Few people would regard the RSPCA as part of the private security industry. Given these distinct differences, the RSPCA could be considered as in the grey area at the margins of the private security industry.

Thus, even with the above test, such is the complexity of the private security industry and related activities, of organisations at the margins and of public organisations engaged in policing, that there are still difficulties in drawing the line. However, the test does go some way towards identifying those parts of the public sector that can be considered as private security and those others at its margins.

Occupations with significant security roles

The third problem in defining the boundaries of private security is that there are many employees who perform security functions as part of their general duties, but for whom these functions are not their main role. A receptionist whose main job is to greet individuals and deal with enquires may also have an access control function and perhaps even a CCTV screen to monitor. Would one consider this person to be working in private security? Again, ordinary staff such as cleaners may be used to search for suspicious items if there has been a bomb scare. Does this mean that a cleaner who has to do this regularly is part of the private security industry? Many citizens also perform voluntary duties which could be construed as security functions, for example, neighbourhood watch schemes and voluntary street patrols. Would these activities be construed as part of the security industry?

This problem can addressed by including within the bounds of private security only those people whose occupation, whether full-time or part-time, consists primarily of a security role, and where there is an employment relationship, whether as employees or self-employed. Thus the receptionist whose primary duties are dealing with enquires would not be included because the security functions would not be his/her primary role. Those who perform voluntary security duties would also be excluded, as there is no employment relationship. However, those whose security role is significant, though still not their primary duties, could be considered on the margins of private security.

Cultural differences

Another factor in drawing the boundaries of the private security industry is cultural differences between countries. What is considered as a private security product or service in one country may not be in another. One of the most striking examples is that of firearms in the USA. These are often marketed as personal security products, and many security officers are also routinely armed. Thus, where in the UK a woman who feels vulnerable might carry a personal attack alarm, in the USA another option might be to purchase a handgun. In the USA certain firearms can be seen as a form of private security whereas in the UK, because of the restrictions on their private use, this could never be a legitimate option. Consequently there are also cross-cultural problems in drawing the lines where private security begins and ends. To address this from a global perspective would be very complex indeed; for the purposes of this book the definition of private security will therefore be examined from a British point of view.

Conclusion

The discussion in this chapter led us to an attempt to identify some functions that might define private security. This is not intended to be the definitive definition, nor do we claim it as such. Rather, we hope the discussion in this chapter, and our attempt at a collection of words that captures the essence of private security, will at least provide a basis for further debate.

The term 'private security industry' is a generic term used to describe an amalgam of distinct industries and professions bound together by a number of functions, including crime prevention, order maintenance, loss reduction and protection; but these functions are neither common nor exclusive to all the activities of the private security industry, though the more that apply to a particular activity the more clearly it can be considered as private security. To be included within the industry, personnel must have a primarily security role, whether this is full-time or part-time, and there must also be an employment relationship, whether as employee or self-employed. The industry also includes certain public sector security employees, where their role is paralleled in the private security industry; the interest served is private and they hold no special statutory powers. The services and products of the private security industry are also generally categorised into a number of distinct sectors. There is also a large grey area around these sectors — and between them, in some cases — which we have called the 'margins' of the private security industry. These sectors will be defined and explored in more detail later in this book, where it will be shown that some are themselves difficult to define. There are also cultural differences between countries as to what constitutes private security. Ultimately, no watertight definition will ever be found for private security unless preconceptions of what it includes are changed. At the very least, however, this chapter should have offered a basis from which a more rigorous debate can commence.

Chapter 3

The Origins of Private Policing and Private Security

Before the beginning of the modern police in England in 1829, the system of policing that existed was largely private and voluntary (Emsley, 1991). In that year the formation of the Metropolitan Police (by Sir Robert Peel) led to a period when policing became increasingly associated with the state police. Over the last 50 years, however, there has been a substantial growth of private sector involvement in policing, leading Johnston (1992a) to call it a 'rebirth of private policing'. Despite the debate over whether this was in fact a rebirth, what is not in doubt is that this latter period has seen the emergence of a substantial private security industry. Chapter 2 illustrated the wide range of products and services that this industry encompasses today. Many of these, however, have antecedents in a multiplicity of occupations and artifacts throughout history. In short, the private security industry is a relatively new phenomenon, but many of its occupations and products have a much longer history. This chapter will explore the origins of the modern industry. It will do so under four distinct periods: private security in ancient times; policing, security and punishment up to 1829; 1829 to 1945, when the 'modern police' became first established; and finally the rise of the modern private security industry, from 1945 onwards. The focus of this chapter will be largely on England, although some references will be made to other countries where appropriate. This chapter is not intended to be the definitive account of the origins of the private security industry, rather to be illustrative. It is our aim to produce a more complete history in a later book.

Private security in ancient times

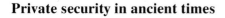

'The locksmith profession is claimed to be the second oldest in man's history' (Master Locksmiths Association promotional literature).

Many functions and crafts that were carried out in ancient times — examples can be found of individuals undertaking lock and key making, investigations and guarding — are today the province of organisations and individuals constituting the private security industry.

As soon as man claimed private property, others sought to steal it; hence with its birth came the need to protect it. To achieve this physical devices were created, and ultimately people employed, to protect private property. Perhaps the first such physical devices were locks. Mention of them can be found in the Bible and Homer, while some of the earliest locks have been found on the ancient site of the great palace of Sargon at Khorsabad, 20 miles north of Nineveh. Designs for locks were also found drawn on the frescoes of the Temple of Karnak, on the Nile. The making of locks could be some 4,000 years old (Monk, 1987), and one of the first security vaults can be traced to a treasure temple built by Rameses III of Egypt, 3,000 years ago (Byrne, (1991).

Locks and keys were also made throughout the period of the Greek and Roman empires. Excavations in Pompeii revealed the house of a locksmith, buried by the eruption of Vesuvius in AD 64, containing lock parts, padlocks and keys; and such was the enormous size of keys in Ancient Greece that wealthy owners would often employ a slave to carry them (Monk, 1987). In Britain, locks can be traced to the Roman occupation and throughout Norman times.

There are also examples of the use of alarms in ancient times, at least of a living kind. From the very earliest time man has used dogs to guard property and to bark at the sight or sound of intruders. Another example is the geese that were said to have saved Rome from the impending attack of the Gauls in 390 BC. Even today one producer of Scotch whisky uses geese to help protect 25 million gallons of spirit (Byrne 1991)!

However, vaults, safes and locks were usually the preserve of the rulers and the rich. The humble usually had little worth stealing and lived in small communities where thefts of others' property would be difficult to hide. Consequently there was little demand for locks and keys amongst the vast majority of citizens. Locksmiths were in these early times a tiny band of craftsmen who practiced their craft for a small elite. It was not until the industrial revolution in Great Britain that the craft was transformed into a mass industry, catering for the whole of society.

Private investigators can also be found in ancient times. At the time of the Phoenicians there was an example of an individual spying on another for gain — by obtaining information for a member of the royal household that the latter's wife was having an affair with the leader of the army. There are also examples of spies in the Bible, such as Joshua and his eleven companions who were sent to spy on the people of Canaan (Draper, 1978) — though perhaps a comparison with modern spies would be more appropriate for this example. There are also numerous examples of property, valuables and people being guarded in ancient times, though it was usually the military or some other state body that performed the guarding — for instance, the personal (Praetorian) guards of the Roman Emperors.

Policing, security and punishment up to 1829

This period saw the emergence of more distinct security roles. Occupations began to emerge that can be more easily linked to and identified with the modern private security industry. Alongside these developments the system of policing and punishment which began to evolve was often private, voluntary or both. This section will explore some of the many examples from this period that illustrate the antecedents of the modern private security industry.

The role of the state in policing and security

Modern policing can trace its lineage back to the tribal laws and customs of the Danish and Anglo-Saxon invaders (Critchley, (1978). The systems of governance, policing and military organisation established by the Anglo-Saxon kingdoms in England contained a number of principles, many of which were not expunged by the Anglo-Saxons' defeat at Hastings. The King was at the apex of all three of these systems, and as Melville-Lee (1901:7) pointed out a century ago:

> The Military and Police systems were closely allied; the national militia was organised in tythings and hundreds, and had a place to fill in the complete design of peace maintenance ...

Administration was divided into tythings, each of ten households, with a headborough at its apex. Ten tythings were grouped into a hundred, each of which was led by a 'hundred man' or royal 'reeve' and, at a higher level, the 'shire reeve' (sheriff) was answerable to the king for his shire. Members of these groupings were responsible for self-policing. They could be called out by the 'hue and cry' and organised in pursuit of a criminal under the 'posse comitatus' (the power of the county). This local system of collective security, with reciprocal obligations to the king, was known as the 'Frankpledge'.

The Norman and medieval periods proved turbulent. Yet despite the instability caused by dynastic struggles and economic pressures, the inherited 'policing' system endured, and was even marginally improved. The Conqueror did not begin *ab initio*. 'Frankpledge', 'hue and cry', 'posse comitatus', sheriffs and much else survived in modified form, and new roles such as that of constable and the justice of the peace emerged. This period witnessed the establishment of a legal framework that laid the foundations of English policing for the next six centuries.

Numerous grants of charters to towns and cities established the general principles of policing and law and order in the lands conquered by the Normans. The *Statute of Winchester* was probably the most significant, and confirmed the concept of towns having watchmen, the building of the town walls, and the re-establishment of the 'hue and cry'. Critchley (1978:7) wrote that:

> [the Statute of Winchester] was the only general public measure of any consequence enacted to regulate the policing of the country between the Norman Conquest and the Metropolitan Police Act 1829, so that for nearly 600 years it laid down the principles.

The Crown was also historically a major builder. Many of its constructions had significant protective measures that are comparable to many modern private security products or strategies. During the Anglo-Saxon period major fortifications included the construction of the burhs, with Offa's Dyke being perhaps the most famous example. The Normans also embarked on massive castle building programmes to provide protection against external foes, to maintain public order and to reaffirm royal supremacy.

The Crown also employed thousands of watchmen up to and beyond the nineteenth century. They were very widely used to provide protection for Crown property. They were generally poorly paid and distinguished by a variety of other titles, including warehousemen, porters, gatekeepers, doorkeepers, doormen, overseers, storekeepers, officers of the water guard, messengers and searchers.

Hybrid policing, security and punishment

Before the evolution of the state into what would be understood as its modern conception, differing institutions and agencies undertook 'public' functions, such as providing jails and military personnel. Despite having a linkage with the government or the Crown, these entities relied to a significant degree on private sources of funding or management, in essence an early form of privatisation of policing, security and punishment.

Policing responsibilities in England were overseen not by the Crown but, in the main, by local government, or rather sub-state government. Magistrates, constables and watchmen, although theoretically appointed by the state, were very close to their respective communities and operated quasi-independently of the state. Policing arrangements from Exeter to York depended very much on local factors.

We are so used to regarding the structure of sub-state government in England as weak that we often forget how significant in our history have been manorial, parish, county, town and city government, and their respective responsibilities for policing. There is a remarkable continuity of eleventh-century local government through most of the following millennium. Jewell (1972) describes the structures during the Middle Ages as 'very much a local affair'. Little contact was experienced with the central structures of the state, other than by key officials such as the sheriff and magistrates.

Policing was almost exclusively a local responsibility; local forces could be raised for wider military and policing purposes but in normal circumstances it was the constable, watchmen, sheriff, 'hue and cry' and 'posse comitatus' that were relied upon. Sheriffs and local magistrates were an arm of the Crown, but they were still rooted in their localities and were not always compliant. Sub-state government in England (unlike many other countries) was quite resilient and semi-autonomous; towns had numerous charters conferred on them by the Crown and enjoyed a considerable degree of self-government. Those who were elected to the roles of 'constable' and 'watchman' also often paid proxies to undertake their duties — another example of early 'privatisation'.

The Bow Street Runners' horse and foot patrols provide another interesting insight into a form of organised 'hybrid' policing organisation, pre-dating the Peel era. Founded in the eighteenth century and based in London under the control of the Bow Street Magistrates, they were essentially a private force and prone to corruption. This small band of men would sell their 'policing' services to government departments and, for a fee, travel to the provinces to investigate. Many of them left significant fortunes behind them when they died. The Bow Street Runners had a mixed press, Radzinowicz (1956:263) calling them:

> a closely knit caste of speculators in the detection of crime, self-seeking and unscrupulous, but also daring and efficient when daring and efficiency coincided with their private interest.

The Thames River Police provide an example of a private force, which eventually became integrated into public policing. The loss of cargo in the Thames near London was astronomically high in the eighteenth century but Patrick Colquhoun and John Harriott persuaded the West India Company that this could be reduced. The scheme proposed was the establishment of the Marine Police Establishment in 1798; the results were spectacular. A few years later this private force came under public control when the Commons approved the setting up of the Thames Police Office, although initially the constables were still paid by the companies. This was therefore a 'hybrid' force, which received financial and other support from the government, while its preventive department (one of four) was paid for exclusively by West India merchants.

The army had a role in maintaining public order during the Tudor, Stuart and Hanoverian periods, but this was often undermined by failings in the quality of the troops and integrity of the officers. Many of the infantry regiments were virtually owned by their colonels, who received payment for their regiment from the Crown, and officers could purchase their commission. In the absence of a French-type police force, the army was often used to suppress insurrection and combat radical agitation. The army or militia was frequently deployed to carry out basic policing tasks and to maintain the peace during public order crises such as food riots and industrial or agricultural disputes, and even during the controversies aroused by electoral and parliamentary reform. The use of the military in policing duties could be unpopular, costly and unreliable. One of the worst crises in civil — military relations was the Peterloo massacre of 1819 in Manchester. Mounted yeomanry were sent by the magistrates to arrest the renowned orator Henry Hunt. They charged a vast crowd, killing 11 civilians and wounding about 400. Incidents such as 'Peterloo' underlined the need for a non-military solution to the growing public order problem — namely, a police force.

The early Royal Mail system also provides us with an insight into instances where state functions relied upon private elements. Before the emergence of the railway network the Royal Mail carried mail in stagecoaches. The mail-coach, horses and driver were owned privately by contractors, yet the security and management were the responsibility of the Royal Mail, which appointed its own armed guards (almost an early form of the private finance initiative!). Travel could be a dangerous experience, threatened by highwaymen and footpads. The mail-guards were liveried, well-trained and well-paid, often ex-servicemen, armed with blunderbusses and pistols which they were prepared to use. The Royal Mail also developed its own in-house investigations department; the Chief Archivist of the Post Office maintains that the Post Office Investigation Department goes back over two hundred years, making it one of the oldest investigative forces in the world!

The prison system in English history, although technically belonging to the state, had a strong private flavour up to the nineteenth century. The system was characterised by fragmented ownership and management, united only by squalor, harshness and private enterprise. At first sight, English prisons would appear to have been very clearly within the responsibility of the state. As Webb and Webb (1922) argued, in English law the prison had always belonged to the King, yet up to 1877 most prisons had come under the responsibility of local government, magistrates, the ecclesiastical authorities, the feudal aristocracy or private individuals; and prisons were even built and managed for profit. Much has been written about the purchase of the right to run a jail; this was expensive, but the charging of fees and the sale of luxuries and indeed essentials and even release, ensured a generally good return on the investment (Ignatieff, 1978).

The reformer William Cobbett, who was imprisoned from 1810 to 1812, was fortunate in having wealthy friends and supporters, which permitted him rather luxurious facilities in the governor's private apartments. He was able to continue his business activities and his writing, while receiving his wife and children on a regular basis. Prisons were subject to government inspection only after 1835. However, it was not until the Prisons Act of 1865 that the Prison Commission was established to create a national system, which it achieved largely by closing down many locally controlled prisons, whose management was invariably corrupt (James, Bottomley, Liebling and Clare, 1997).

Private policing and security

The limited role of the state in policing and security meant the majority of the functions were undertaken by what might now be called non-state, sub-state or private institutions, or by groups of individuals.

In the absence of a national, efficient police force, various 'initiatives' flourished to fill the void. One of these private endeavours included approaching 'thief-takers', such as the rogue Jonathan Wild, who would hunt thieves down for financial gain. Individuals or groups might join an association for the prosecution of felons. For the payment of an annual subscription, the association would undertake investigation and prosecution; some maintained patrols and hired watchmen. Within this spirit of initiative it must be noted that investigations into crime were usually undertaken by the victim. Some degree of co-operation with constables was called upon, although they themselves rarely investigated the crimes.

The lack of professional investigators meant private investigations were commonplace. The task could also be undertaken by enthusiastic magistrates or ordinary employees. For instance, in the eighteenth and nineteenth centuries the Royal Bank of Scotland dispatched bank clerks to undertake investigations and pursue alleged criminals. William Spurrier, in one example from the late eighteenth and early nineteenth centuries, was engaged from his law office in Birmingham in the detection, apprehension and prosecution of forgers and counterfeiters on behalf of the Bank of England (Cook, 1995). Thus the practice of catching thieves was effectively a freelance role. If one was robbed, the options were to suffer, to advertise for the return of the stolen property, or to approach a thief-taker direct.

In the absence of local policing, citizens often intervened directly in neighbourhood disputes and crime prevention. There were numerous instances of public initiative, which might occasionally descend into vigilantism or intimidation in the form of 'rough music'. The efficacy of such policing was more suited to a parochial rural society than to an urbanised area with an expanding population, and the effectiveness of such traditions and sanctions was curtailed by the radical changes wrought by the industrial revolution.

In rural areas the battle to deter poaching saw the local gentry employing forms of private security to put intruders to flight. Crime in the countryside included the theft of animals and birds owned by the Crown or landowners. The motives of the poachers were highly varied. Retainers and gamekeepers were hired by the feudal nobility and landowners in the Middle Ages to protect their property. The battle to curtail poaching has been likened to a military conflict, with the landowners and village communities forming opposing armies. The number of gamekeepers grew throughout the nineteenth century, though the poaching wars declined in intensity and violence. Gamekeepers were often licensed by the bench and brought the guilty before magistrates. They had police-like powers of arrest and could search without warrant. Unsurprisingly gamekeepers were unpopular; nevertheless, anything that resembled a foreign (ie, centralised) police force at this time was regarded as suspect. Private watches also existed and were reasonably well-run by the authorities in places such as the City of London and Westminster. In the late eighteenth century improvement legislation also enabled local administrations to enhance their watching arrangements.

Medieval guilds were also commonplace across Europe. They represented crafts and trades and were powerful bodies. They could discipline their members and had an influence over law and order in their respective locality. For instance, the Guild of Saddlers, which originated in the eleventh century, had strong disciplinary powers over its members, including the ability to commit them to prison for debt, for other, fairly trivial offences and also for unacceptable professional standards.

Policing, security and punishment 1829–1945

In retrospect, the period up to and beyond the formation of the modern police can best be described as a 'mixed economy of policing' (Gill and Mawby, 1990). Between 1829 and 1945 the 'new police' gradually expanded and became the predominant policing body. However, the 'hybrid' and 'private' aspects of the policing system did not die out but simply became increasingly overshadowed. This section will illustrate the continuing 'private' and 'hybrid' contributions to policing and security.

The gradual emergence of the 'new police'

The 1829 reforms were by no means the first legislative attempts to reshape policing. The view that 1829 is *the* watershed in modern British policing is a moot point. The imperfections of the pre-reformed system of policing and security were not taken for granted by contemporaries; indeed, many eulogised the system even after it had been largely supplanted. Reynolds (1998), for instance, writes that the paid watchmen in certain central London areas 'were tantamount to professional law-enforcement agencies'. She argues that 1829 was not a radical departure, as has so long been maintained. However, as Wall (1998: 19) comments:

> By today's standards, however, it would be wrong to call them police, they were more a cross between security guards and bounty hunters. The detection of crime was over and above their ordinary duties.

Indeed, it is evident that Peel's reforms took decades to be implemented. The pre-reform system was not expunged even in London in 1829 and it survived elsewhere for decades. The 'old system' thus survived and prospered for a long time after the rise of the new police.

The new policing system was not free from problems, and took many years to 'bed down' and to extend across the nation. Of direct relevance to our study is the clear fact that this nascent system actually co-existed with the traditional genre of watching for a considerable period. It was the *County and Borough Police Act* 1856 that signified the end of parochial control of local policing, although perhaps its importance has been overshadowed by the 1829 Act. Taylor (1997) claims that this latter Act laid the foundations for the system of policing that we have today, but it too took some time to become truly effective. As Emsley (1991: 43) has written: 'As the 1840s dawned there was no single model of policing dominant in England'.

Most of the boroughs had implemented the policing requirements of the *Municipal Corporations Act* by the end of the 1830s, but as late as 1856 there were at least six, and possibly as many as 23, who had not. Critchley (1978:58) saw the thirty years following the 1829 Act as a 'series of somewhat hesitant steps'. The success of the London model was not guaranteed. Resistance to the new methods was not hard to detect, as Storch (1975: 216) quips:

> By the 1830s, then, there was no shortage of unreconstructed small conservatives who wished to maintain the old arrangements, but they were declining in number. There were at the same time very few indeed who supported the idea of a centralised national police force ...

Hybrid policing and security

The nineteenth century also witnessed the rise of specific police forces in dedicated sectors of industry. These were quasi-public, in the sense that their role and functions were prescribed by legislation. The *Special Constables Act* 1831 enabled railway companies to deploy special constables. Other legislation empowered railway companies to appoint their own special constables, who ultimately became in-house police forces appointed by each of the private companies, and were dressed similarly to the Metropolitan Police. The *Harbours, Docks and Piers Acts* 1847 provided for owners to appoint special constables (some of who still exist!). In certain instances private/public policing boundaries were not clear. For instance, the policing of the port of Dover from 1836 to 1933 was undertaken by the Dover Borough Police although the cost was born by the Dover Harbour Board; thus even the new police could be hired. Glamorgan Constabulary provided a regular police officer for a public corporation, the National Coal Board, on a contractual basis (Bruce George's father, Police Sergeant George, was one of these officers!). There were also dedicated police forces in numerous other areas including markets, universities, parks and canals.

Private policing and security

Early forms of private security could also be found in many areas of Victorian life, particularly entertainment. Many a pub, from the eighteenth century onwards, had large imposing men ready to bar admission or eject the unruly (antecedents of 'bouncers'); many were ex-pugilists as were many landlords. Violence at Doncaster races also frequently occurred and necessitated firm measures. To combat these disturbances the racecourse employed an ex-policeman, and gatekeepers to deal with breaches of law and order. Indeed, at the Doncaster St Leger meeting in 1829 armed roughs marched on the course, but were seen off by a group of mounted hunting men armed with whips and followed up by dragoons, militia, yeomanry and special constables. The charge that put the marchers to flight included grooms and other employees. At theatres, policing and security was based more on vigilantism; at the Globe in London, apprehended cut-purses and other criminals were manhandled on to the stage, where they were temporarily enchained and abused.

The wealthy elites employed their own methods of securing their property and ensuring their personal safety. The aristocracy hired vast numbers of servants to provide protection inside the buildings, and when they were travelling outside. Gamekeepers were charged with the duty of protecting their livestock, and coachmen provided security on their journeys. The steward, butler and housekeepers supervised and disciplined the staff and oversaw security, which included ensuring the numerous doors and windows were locked and that valuable property was secure. Underneath them was a range of other staff, usually deferential but also sometimes disloyal, insubordinate and criminal. In Linebaugh's (1991: 366) words: '... the hierarchy of servants within households corresponded to their security responsibilities'.

The history of the English industrial revolution has been well-documented. It is not, perhaps, so well known that within industry a significant contribution to the emergence of the modern private security industry can be discerned. Not only did factories have their own formal and informal methods of security and discipline, but it was the exponential pace of technological development that facilitated the rise of firms like Chubb, manufacturing such security hardware as locks. The growth of large enterprises, with hundreds or thousands of employees engaged in mining, shipbuilding, the railways and engineering in urban centres, meant that the old night watchmen and social control gave way to new models of private policing. This saw the growth of works police and in-house private security. In some traditional industries, notably iron and steel and coal, the local constables had their duties extended to patrolling private premises.

The growth of factory systems in the eighteenth and nineteenth century saw growing anxiety, on the part of owners, about security. Parliament was especially active in passing laws on theft and other offences in different sectors of industry, but without proper means of enforcement. Within factories discipline could be exercised by the foremen (or 'butties' and 'doggies' in mines), or by the more enlightened methods which the philanthropic entrepreneur Robert Owen used in his New Lanark factory. Naturally, the threat of disciplinary action or dismissal could prove effective in ensuring good behaviour; occasionally firms also used a system of fines to deter unwarranted action.

Private investigators also operated in Victorian England. The famous Pinkerton Detective Agency collaborated closely with the Metropolitan Police in pursuit of Adam Worth — 'The Napoleon of Crime' — whose many heists throughout the world included the theft of one of England's most famous paintings, Gainsborough's 'Duchess of Devonshire', from Agnew's Art Gallery in 1876. All these non-state 'forces' were in place by the 1880s, running parallel to the official police.

The threat to bank security included robbery of bank premises, theft of cash or valuables in transit, forgery of notes and coins, and internal fraud. From the earliest times banks had to protect themselves, with the support of the local constabulary. The banks had safes, strong boxes and wrought iron chests, and security was part of the general duties of bank officers; there was little recourse to outside companies for assistance with investigations. The banks had a near monopoly on the commercial storing of money and valuables, though in 1882 the Chancery Lane Safe Deposit was opened, renting strong rooms for the storage and safekeeping of silver, documents, jewellery, etc. Some of these would be deposited overnight by nearby Hatton Garden shops. These vaults or strongholds were so secure as to have defied not only all attempts of robbery but even the Luftwaffe's attentions.

The manufacture of security hardware became an enterprise of some magnitude in the nineteenth century. The foundations were laid for those modern companies which are integral to the present architecture of the private security industry. Famous concerns include Chubb, which started in the hub of lock-making, and Willenhall, Staffordshire. At the end of the century the latter was selling locks, alarms (influenced by American innovations) and chests to banks and private individuals. The company developed a presence throughout the British Empire.

The Corps of Commissionaires was founded in 1859, supplying businesses with commissionaires, who often undertook security functions such as guarding (Reese, 1986). As the Corps still provides such services today, it is probably the oldest private security company in the UK. Towards the end of this period, in 1926, Arnold Kuzler also founded MAT Transport, which was probably the first 'modern' cash-in-transit security firm (Draper, 1978).

By the end of the nineteenth century, many of the building blocks of the modern private security industry were in place. Within commercial, industrial and government departments in-house security had become firmly embedded. As previously alluded to, ports and railways had established private police forces, albeit sanctioned by statute. Security hardware was also becoming a profitable business of some magnitude.

The rise of the modern private security industry, 1945 to the present day

The period after the end of the Second World War witnessed the emergence of the modern commercial private security industry. Many of the roles and activities that were previously undertaken by 'ordinary' employees became increasingly the preserve of specialist in-house or contracted security personnel. In other areas, the changing risks faced by individuals or organisations led to the purchase of security products and services. These services and products were increasingly provided by specialist security companies. Thus what distinguishes this period is the emergence of specialist firms dedicated to providing security services and/or products. One of the first contract guarding firms is believed to be Night Watch Services, which was founded in 1935 by Edward Shortt. The company was not initially successful, and just as it was about to fold two clients, Henry Tiarks and the Marquis of Willingdon, took over Night Watch, changing the name to Night Guards Ltd. The company was reformed after the Second

World War, becoming Security Corps in the early 1950s and then Securicor in 1953 (Underwood, 1993). Hence the establishment of this firm marked the beginning of the first 'modern' security companies, as Securicor still exists and indeed is one of the largest companies in the UK.

The contract security industry started to grow significantly during the 1950s. Securicor finally began to flourish and the Corps of Commissionaires still offered its members for security functions. In 1951 the forerunner to Group 4 was established, with the creation of the guarding company Plant Protection Ltd, in Macclesfield. By 1958 it had over 200 employees and was undertaking its first cash-in-transit work. The name of the company was changed to Factory Guards in 1963, and the corporate member companies Store Detectives Ltd and Securitas Alarms were founded. In 1968 the company's name was changed yet again, to the famous Group 4 Total Security Ltd, and the company has since grown into one of the largest in the UK, if not the world. In the same year Chubb Wardens was founded, marking Chubb's entry into the guarding sector.

During the 1950s the use of burglar alarms also became widespread, stimulated by rising crime and the requirements of insurers. Chubb established itself as a major player in this market when, in 1959, it bought a third share in Burgot Alarms; and in 1960 Burgot bought three other significant alarm companies. The growth of this sector is illustrated by the increasing number of intruder alarm companies. In 1960 there were 11 known companies, which by 1970 had grown to the 98 listed in the trade directory; and this was probably an underestimate (South, 1988). Similarly, in 1971 Chubb Alarm's turnover was £6.5 million, where Burgot's had been £80,000 in 1948.

In 1950 it was estimated that the private security industry had sales amounting to £5 million per year; by 1967 Clayton's (1967) description of the three largest companies in one sub-sector illustrates the remarkable growth of the industry. For instance, in the cash-in-transit sub-sector the largest company was Securicor, with nearly 90 branches, 600 vehicles and 6,000 uniformed employees, followed by Security Express with 250 vehicles and 1,200 employees, and by Factoryguards, also with around 1,200 employees, operating in a sector reputed at the time to be worth some £50 million. Meanwhile, the safe and burglar alarm manufacturers were estimated to have a turnover of some £40 to £50 million, employing over 30,000 individuals. By the late 1960s Securicor had over 18,000 employees and 3,000 armoured vehicles and provided a plethora of services, from patrol services, cash-in-transit and courier services, to security advice (Bunyan, 1976). By the mid-1970s the industry had grown to such an extent, that there was wide divergence over the numbers employed, from the 250,000 estimated by Bunyan (1976) to the 100,000 by South (1988). Such diversity in estimates continues today, but, whatever its true size, the modern industry has undergone dramatic growth from the 15 officers on bicycles who started with Night Watch Services in 1935.

Conclusion

In this chapter we have shown that the origins of the private security industry can be traced back thousands of years. We have illustrated the many roles that acted as antecedents to the modern private security industry, and have also demonstrated the wide extent of private involvement in policing, security and punishment throughout the ages, including the period following the formation of the 'modern' police in 1829. Finally, we have shown the metamorphosis of a small peripheral service sector for the rich, into a multi-million pound industry by the end of the twentieth century. The evolution of the private security industry resembles the branching of a tree as it grows: for, as the industry has developed over the last 50 years, so its range of services and products has branched out and grown. The many branches of the modern industry will be explored in the second part of this book. Given the huge growth of the industry, however, it would be useful first to explore some of the reasons for this growth.

Chapter 4

The Growth Of Private Security

The previous chapter illustrated how the modern private security industry has been transformed from a few small and diverse companies in the early 1950s, to a multibillion-pound industry employing hundreds of thousands in the 1990s. Whole sectors of private security have grown from virtually nothing into huge industries in the space of 40 years. The industry continues to grow, even in adverse economic conditions, leading some to call it 'recession resistant' (Williams, George and MacLennan, 1984). This chapter will examine the size of the industry, then the growth of private security in the post-war period, and will offer some explanations of the causes of this huge growth.

The size and growth of the private security industry

The size of the private security industry

It clearly follows that if the industry itself is difficult to define, then its size will also be difficult to estimate. Few of the main estimates of the industry's size take into account sectors such as door supervisors, stewards, private-sector detention services, or security storage and shredding services. Furthermore, certain security product sectors such as detection equipment, security glazing and fences are also frequently excluded. Research by the British Security Industry Association (BSIA, 1994b), Jones and Newburn (1994), Jordans and Son (1992) and Key Note (1993) provides notable examples of this trend. The other major problem is that many estimates are not based on empirical research. For instance, Bunyan (1976) seemingly plucked the figure of 250,000 from the air when estimating the numbers employed in the industry. We have therefore attempted to gauge the size of the industry by combining estimates from different sub-sectors. Some of these are based on empirical research of varying quality, others are the estimates of experts, and some are based on our own rough calculations. It must also be stated that unfortunately, in a number of sectors — because of the difficulty in gaining the information needed — some of the figures are dated, and estimates are used from differing years to arrive at a total for the industry. Explanations of the source of the figures and how they were arrived at can be reviewed in the relevant chapter, where each sector and its size is considered in more detail.

We have erred on the side of caution with all the calculations and thus they probably represent an underestimate. At the very least, what Table 2 (see page 30) shows is that when sectors normally excluded from estimates of the industry's size are included, the total size is substantial. Most current estimates of the entire industry remain unrepresentative because they fail to take account of these sectors. In the future, more considered research on this subject should attempt to undertake calculations based on the wider industry. We accept there are flaws in our methodology but the aim was not to arrive at an accurate figure for the industry's size, but rather an estimate of how large it might be when the wider industry is included. Perhaps not until there is statutory regulation will more rigorous and accurate figures be available, and perhaps not even then.

Table 2. The size of the UK private security industry

Sector	Total employees	Turnover (£m)
Static guarding	125,000	1,470
Cash-in-transit	15,000	382
Door supervisors	50,000	150
Stewards	12,000	12
Close protection	1,000	10
Detention services	6,000	150
Security storage/shredding	2,500	100
Security consultants	1,000	20
Professional investigators	15,000	225
Security products	90,000	2,961
Total	**317,500**	**5,480**

Sources: for static guarding, BSIA (1999b), CIT, BSIA (1999a) and interviews with experts; for door supervisors, GMB (cited in *The Guardian*, 15 April 1996) and informed estimate; for stewards, Football Trust Survey (cited in *The Observer*, 11 January 1998) and informed estimate; for close protection, interviews with experts; for detention services, informed estimate based on various reports; for security storage/shredding, informed estimate based on various reports; for security consultants, interviews with experts; for professional investigators, Institute of Professional Investigators (IPI) (1992) and informed estimate; for security products, see chapter 10.

Note: for more information on the above figures, see the relevant chapter.

The growth of the private security industry

It is clear that the private security industry has experienced substantial growth, particularly if one compares the size of the industry in 1950 with that in 1995. In 1950 there were a dozen or so companies involved in the provision of security services and products; now there are thousands, in what could be described as a multi-billion pound industry. The problems occur when the growth of private security over a shorter period of time, such as the last ten years, needs to be analysed. We have found a number of indicators that illustrate the growth of private security in the post-war period.

Probably the most systematic research into the growth of private security is based on the census and labour force survey utilised by Jones and Newburn (1998). They found that private security employment had increased by 240 per cent since 1951, with the most rapid periods of growth occurring during the 1950s and 1960s. They also discovered that employment had increased by a further fifth since 1971, and even speculate that the industry could now employ as many as 333,000. They base this unsophisticated estimate upon the mean number of employees per firm found in their survey, times the estimated total number of security firms.

Estimates of market size, when comparing studies undertaken by different organisations and individuals, suffer from the same problems as indicators of the total number of employees. This is not a problem when comparing studies by the same organisation or individual, if they are repeated longitudinally and use the same methodology. For instance, Jordans and Son estimated the private security market to be worth £357.5

million in 1981, £581.5 million in 1985 (Jordans and Son, 1987) and £1,225.6 million in 1990 (Jordans and Son, 1992). The Jordans survey illustrates the growth of the private security market, but underestimates its size by not including sole traders, partnerships, in-house security or private investigators.

Reasons for the growth of private security

Many determinants have affected the growth of private security. They can be broadly divided into those that have increased the demand for it, and those that have made expansion easier. Some of the main influences increasing the demand for private security include the growth in and fear of crime, terrorism and environmental protest; the demands of insurance companies; the underfunding of the police, despite greater demands on their resources; and the privatisation of aspects of the criminal justice system. The increase in private property, positive research illustrating the benefits of private security, aggressive marketing of security products and services, and changes in technology have also been influential in fuelling a 'security race'. Another factor that has made expansion of the industry easier, is the unregulated state of the market. However, one can only identify the influences on growth in certain sectors of the industry, and then speculate about their impact. The lack of reliable data regarding the size of the industry as a whole and of particular sectors makes more sophisticated attempts at correlation difficult.

The increasing experience of crime and fear of it

The rise in crime in the post-war period has been the fundamental driving force behind the growth of private security. There is little doubt that crime has risen in the post-war period and a number of indicators can be drawn upon to illustrate this growth. The number of recorded indictable offences in England and Wales has increased dramatically. In 1945 there were only 478,494 such offences (Morris, 1989) but this had risen to 5,100,241 by 1995 (Home Office, 1997) — an increase of more than 1,000 per cent over 50 years! This growth has been typical of most 'Western' industrialised countries, bar Japan, where crime has remained relatively static (Smith, 1996). However, there is much debate over the accuracy of the recorded level of crime in relation to the real level of offences (see Maguire, 1994).

There are a number of surveys that seek to gauge the real level of crime and these also illustrate a general increase in that level throughout the country. The most significant is the British Crime Survey (BCS), which has been taking place annually since 1981 and which surveys domestic households for their experience and perception of crime. Between 1981 and 1995 the BCS reported a 109 per cent increase in acquisitive crime per 10,000 adults, a 9 per cent increase in vandalism and a 17 per cent increase in violence (Mirrlees-Black, Mayhew and Percy, 1996).

Research has also demonstrated the fear of crime. The BCS has found that in 1998, 19 per cent of those surveyed were very worried about burglary, 18 per cent about mugging and 21 per cent about car theft. Eighteen per cent of women felt 'vey unsafe' on the streets at night. (compared to three per cent of men), whilst for women over 60 the figure was 31 per cent. Perhaps the best illustration of the fear of crime in proportion to the real risk was the finding that 58 per cent of respondents thought at least one-half of crimes recorded by the police were violent, when in reality it was only 8 per cent (Barclay and Tavares, 1999).

However, it is important to distinguish between general trends in crime, fear of crime, and more specific trends in particular types of offences. The total number of offences is an aggregate of many different types of crime, which all have a different impact upon the demand for security services and products. For example, if the aggregate trend has increased because of rising car theft, this is likely to have a major impact on the demand for vehicle security products; but such trends will have only a limited impact on the market for domestic intruder alarms, except in creating a general increase in the fear of crime. One must also be careful in linking the growth of specific security strategies to specific crimes, as often a security strategy is employed to counter a wide range of crimes and, often, non-crime-related problems.

Burglary is a crime where there is a clear link with security strategies to combat it. According to the BCS (Mayhew, Maung and Mirrlees-Black, 1993) between 1972 and 1991 there was a 61 per cent increase in burglary. The experience of a burglary, or the perception that one is increasingly likely to occur, may lead to the purchase of security strategies to combat this, such as better locks, intruder alarms, security lighting, etc. Earlier in this chapter the growth in the use of intruder alarms over the last 25 years was illustrated. The increased domestic use of such products has also been illustrated by the BCS (Mayhew et al, 1993). In 1988 the BCS found that eight per cent of respondents had a burglar alarm; this had increased to 21 per cent by 1996 (Mirrlees-Black et al, 1996). Only 35 per cent of respondents had some or all windows fitted with locks in 1988, but this had increased to 69 per cent by 1996. The number of respondents who had some or all doors fitted with double/dead locks increased from 57 per cent to 70 per cent during the same period. The BCS also found that respondents who had experienced a burglary were more likely to take security precautions (Mirrlees-Black et al, 1996).

Unfortunately the BCS does not cover commercial properties, but the same principles apply, if not more so. The threat of burglary to commercial premises is also likely to increase the demand for more commercially related security products and strategies, for instance: grilles on shop fronts, high perimeter fencing, remote signalling intruder alarms, CCTV and, ultimately, the employment of security guards (although the last of these may be part of a wider strategy to prevent a range of risks of which burglary is only one). Similarly, some of the other measures may also have dual uses.

The above example of burglary illustrates how an increase in demand for specific types of security strategy might be influenced. There are many other types of crime that stimulate an increase in certain types of security strategy, but it is not possible to discuss these fully in the space available in this chapter. For instance, increasing concern over vehicle crime has led many to purchase alarms, 'crook-locks', etc. Increasing concern over shoplifting has led many retailers to increase the number of security officers, store detectives, CCTV systems and tagging systems. Concern over robbery may lead to the purchase of increased physical protection, the use of secure transportation, CCTV, etc. A brutal murder or vicious rape may lead to an increase in the use of personal attack alarms by vulnerable members of the community. These provide a brief illustration of how an increase in and greater fear of certain types of crime have a specific impact on certain security strategies.

The rise in terrorism

Terrorism has also had a dramatic impact on the demand for private security. The fear of terrorism creates a market for a range of security services and products. Consequently, expenditure on security to counter terrorism has increased substantially (Latter, 1992). During the 1990 — 91

Gulf War many vulnerable UK organisations increased their security against potential attacks by Middle Eastern terrorist groups. Terrorism did not in fact increase during this period, but there was a greater fear of attack. This is one of the most prominent examples where the perception of a threat is much more important than the experience of an attack. For instance, Cunningham et al (1990) estimated that, given the total expenditure devoted to counter-terrorism overseas by US organisations, $223 million was spent per actual victim of a terrorist attack.

A single high-profile terrorist attack, or the perception that one might occur, can lead to a huge reaction, out of all proportion to the risk. This is because the consequences of just one incident can be so serious. For example, the February 1996 bombing of the London docklands was estimated to have caused damage worth £75 — 150 million (Association of British Insurers (ABIn), 1996). Such a devastating attack, which destroyed whole office blocks, may also have lead to some companies going out of business. There has also been the political imperative of governments not to give in to terrorists.

Some of the reactions of different organisations to a perceived or actual increase in terrorism, impact upon the demand for private security. One of the policies governments might pursue to combat terrorism could be to enact legislation that would require minimum standards of security, as happened with aviation security in the late 1960s and early 1970s. State organisations such as the police and military might bolster their security, which would lead to increases in the demand for security products such as bomb-detection equipment and x-ray machines. It would also result in greater demands for protective equipment such as armour and bullet-proof glass. Organisations and individuals, might also purchase similar security equipment, as well as increasing the number of security officers they use, employing close protection for vulnerable executives and seeking the services of specialists to predict the risk of attack abroad.

The growth of environmental protest

Protest movements can be traced back many hundreds of years (Priestly, 1968). The post-war period was characterised by numerous public demonstrations over a wide range of issues, including industrial disputes, apartheid in South Africa, nuclear weapons and unemployment. Generally, private security employees were not involved with the policing of any of these protests. However, in the 1990s, with the growth of protests over environmental issues and animal rights, private security employees became increasingly involved in policing public demonstrations. Private investigators were also employed to gather intelligence on these protesters, and bailiffs employed to enforce court orders (Button, Brearley and John, 1999).

These protests have been a significant source of income for the private security industry. The costs of security personnel and fencing for the Newbury Bypass, for instance, amounted to £30 million during 1996 — 97 and, at the height of the protests there were over 1,000 security officers employed. The cost of security for the M3 at Twyford Down was £3 million, out of a total budget of £26 million (Bryant, 1996). Other significant road developments have also had to spend increasing amounts on security, whereas 10 years ago security would have been a minimal and negligible cost for this type of development.

The demands of insurance companies

The growth of certain types of crime often leads to increased insurance claims, and these can result in reduced profits for insurance companies. This leads to such companies increasing their premiums or, more importantly, demanding greater security before cover is granted.

Some companies also offer discounts to customers who have certain security devices. It is these latter two factors that have impacted most on the growth of private security. As part of a strategy to combat the increase in claims, many insurers have set minimum security requirements. These vary between different companies, but the Association of British Insurers has produced guidelines setting out the minimum security requirements for domestic properties (ABIn, 1994). Such standards inevitably lead to an increased demand for security products.

The government's commitment to privatisation

The Conservative governments of the 1980s and 1990s were committed to privatisation but in cases where this proved impossible they favoured the application of 'social markets' to public services (Loveday, 1995). These have impacted on the growth of the private security industry. There has been a growing involvement of the private sector in the criminal justice system, and there has also been an increase in the contracting-out of services, including security, in the public sector (Johnston, 1991; Johnston, 1994; and UNISON, u.d.). The most significant source of growth for private security firms has been the prison and prison escort sectors, which will be discussed in chapter 7.

The failure of governments adequately to fund the police

During the 1980s, governments substantially increased expenditure on the police, through higher pay and the recruitment of more officers and civilian support staff (Loveday, 1992). Expenditure on the police increased by 50 per cent in real terms between 1979 and 1989 (Police Foundation and Policy Studies Institute (PF/PSI), 1994), and the actual strength of the police force rose by 5.3 per cent between 1983 and 1993 (Moran and Alexandrou, 1994). However, crime and the demands on police services have risen much faster than police resources and, consequently the police have become more overstretched. Crimes per officer have increased from 26 in 1982 to 42 in 1992 (PF/PSI, 1994). The police also face demands to perform functions that the public hold in high regard, but which operationally have very little impact upon crime. Surveys of the public consistently call for more police officers on foot patrol. In a recent such survey, conducted by MORI for the Audit Commission, over 80 per cent of those surveyed felt 'very' or 'fairly' assured by the sight of police officers patrolling on foot, and three in five stated that they would pay more taxes for increased foot patrols (Audit Commission, 1996).

The consequent reduction in the police's ability to satisfy demand has led to private initiative attempting to fill the vacuum. There are estimated to be over 1,000 private security patrols in housing estates, shopping precincts, hospitals, schools, parks and other private and public facilities. There has also been an emergence of vigilante patrols. For instance, in 1989 a US-model Guardian Angels unit was established to patrol the London Underground, and a 110-man 'Dad's Army' patrol, led by a retired army captain, was organised in Gosforth and Newcastle-upon-Tyne (Bright, 1993). More recently there have been patrols by local residents in the Balsall Heath area of Birmingham, to deter prostitutes and their clients. In areas where there is a demand for a service that the police are unable or unwilling to supply, the private security industry (if there is a profit to be made) and other private initiatives have responded.

The growth of mass private property

A number of academic writers have identified the increase in private property as one of the major causes of the growth of private security. Stenning and Shearing (1981) argue that mass private property is the most important factor in accounting for the industry's spectacular growth; and in a study of the growth of the private security industry in Taiwan, Hou and Sheu (1994) agree. Essentially, the rationale behind these arguments is that, given the rate of growth in private property, there is a greater requirement to protect it, and that this task has fallen to the private sector. The growth of private shopping centres provides a good example. Private security firms have assumed the role of policing these centres, in a manner comparable to the police's patrolling of the old public high streets. The replacement of terraced streets by blocks of flats during the post-war period has also affected the pattern of policing. Private security firms, rather than the police, now patrol the corridors of many private blocks. Some of the country's major convention centres also employ large numbers of private security officers; for example, the National Exhibition Centre, International Convention Centre and National Indoor Arena in Birmingham employ a combined security force of over 160 full-time and over 400 part-time officers.

Government policy towards crime and private security

A stimulant to the private security industry during the 1970s and since then, has been the refocusing of Home Office policy towards crime prevention (South, 1989). Under the leadership of Ronald Clarke, a clear policy shift began towards the end of the 1970s in favour of research by the Home Office Research Unit into crime prevention. The focus altered from developing strategies to change the predisposition of an individual to offend, to analysing the environment so as to devise means to make it less attractive and less easy to commit particular offences (Loveday 1994; and Maguire, 1994). This alteration in focus started a wave of research which illustrated how purchasing physical security, marking property and increasing surveillance could reduce the threat of crime to an individual (Pease, 1994a). Research illustrating the benefit of a device or service is likely to exert a positive influence on some to purchase a specific product. One of the most striking examples of this influence has been the research illustrating the benefits of CCTV, which has been hyped beyond proportion so that it is perceived by many to be a universal panacea for crime reduction; the result has been that the use of CCTV has increased dramatically over the last 10 years.

Governments have stimulated the growth of private security in other ways. One of the more prominent measures taken is the enacting of legislation enabling minimum standards of security to be mandated, as in aviation, maritime and railway security. Such standards cover issues like the screening of luggage, and manual searches, which have implications for the security requirements of an organisation. Governments have also encouraged the growth of the private sector by offering grants for security. The most prominent example in recent years has been the competitions by local authorities for funds to contribute towards CCTV systems. It was estimated by *CCTV Today* in November 1995 that the second competition for CCTV announced in that month could generate expenditure on cameras of £60 million by 1998!

The role of advertising

It can also be argued that the advertising used by private security companies, aided by some government crime prevention campaigns, has contributed to the growth of private security. Many advertisements for security products play on the fear of crime, and exaggerate the likelihood of it occurring. An advertisement for Secustrip and Secubar, for example, pictures a male dressed in black like a stereotypical burglar, with black hat and gloves, trying to force entry through a door, with the caption 'every 20 seconds someone creates a new sales opportunity for our products' (*Professional Security*, June/July 1994).

More recently the government initiated a crime prevention campaign which featured ravens ransacking homes as burglars might. There are many more examples of these types of advertisements, which have arguably helped to stimulate the growth of private security — which is certainly their aim.

Changes in technology

As criminals discover new methods of defeating security systems, so the 'security race' in developing advanced technology security products is fuelled. The 'arms race', as Byrne (1991) has termed it, continues. As organisations and individuals seek the ultimate in protection, and as increasingly complex security devices become available, so ageing systems are replaced and new devices purchased. Organisations are seeking better security to counter the criminals who have successfully mastered existing security systems and strategies. Improvements in technology have strengthened defences against crime — CCTV was once a specialist device installed in only a few locations, but is now deployed in almost every retail outlet, city centre and factory. Not only has the technology improved, but more functions are now available on CCTV at affordable prices, such as better quality picture and recording equipment, colour and night vision; more users therefore upgrade their equipment, fuelling the growth of CCTV systems.

Other factors influencing demand

The health of the construction industry is another key influence on the demand for security products, particularly in the commercial sector. As new factories, office blocks and shops are built, new security systems are fitted, ranging from alarms and CCTV to basic locks. Similarly, the recent growth in the building of prisons has fuelled an increase in the market for products to secure them, such as locks, access control systems and CCTV. Growth in the domestic property sector is likely to increase the demand for simple security devices such as locks. Increased vehicle sales may also fuel the purchase of security products used on cars and lorries.

The deregulated nature of the market

It will be shown later in this book that in almost every sector of the industry there are no legislative requirements on those entering the market, and in certain sectors there is very little capital required. There is also a temptation for consumers to buy the cheapest security service or product, because the purchase is often a 'grudge' cost. It is therefore possible to establish a security firm in some sectors of the industry with no prior knowledge, experience or training, and to gain enough business to make a living. There is also a perception that large amounts of money can be made. Thus, if there is an increase in the demand for security the market can respond very quickly. Another influential factor on the growth of the industry is the high level of unemployment in other sectors of the economy. Many individuals have been made redundant with large lump-sum payments (military personnel and miners, for instance) and some have invested the money in their own security company.

Conclusion

This chapter has illustrated some of the many determinants that have contributed to the growth of the private security industry in the UK and elsewhere. The complex relationship between these various factors makes it difficult to judge the weight of their individual influence. Nonetheless, it is clear to us that the most significant features of that growth are the increase in and fear of crime/terrorism, combined with a generally more affluent society that wants to protect itself. The policy of privatising aspects of the criminal justice system has also been significant. The other determinants — although important influences — would not have mattered as much if levels of crime and terrorism had been declining. Ultimately, the future prospects of the private security industry are linked most significantly to these factors, and as long as these influences remain the private security industry will continue to prosper.

Part 2. A Fragmented Industry: The Organisations, Services and Products of the Private Security Industry

Chapter 5

Organisations

The diversity of services and products offered by the private security industry will be considered in greater depth in the following chapters. However, it is important first to discuss the role and work of some of the many organisations that represent, devise standards for, and voluntarily regulate individuals and firms operating in the industry. These organisations are central to any meaningful analysis of the characteristics and structure of the private security industry. This chapter illustrates the increasing fragmentation of the industry, with a mushrooming number of organisations being established. Fragmentation has also resulted in greater competition, with more than one organisation seeking to represent or regulate a sector.

There are also many organisations external to the industry that have an interest in private security. There are numerous non-security trade associations, such as the British Bankers Association, with a security committee. Constraints on space preclude a comprehensive analysis of these, since it would probably encompass most such associations in the UK. The focus of this chapter, therefore, will be on those organisations that directly represent, devise standards for, or voluntarily regulate some aspect of the UK private security industry. We will also briefly explore some of the growing number of European and international organisations that are beginning to emerge.

The organisations within the industry range from large companies with many employees and million-pound budgets to associations run by dedicated, unpaid individuals in a back bedroom or from their place of work. Some are well known, and seek publicity wherever possible; others are more secretive, and it is difficult to gain information about their activities. There are also some informal gatherings of security professionals, which need to be considered. The organisations relevant to the industry can be distinguished by a number of criteria: first, there is a distinction to be made between national and international bodies; second, some organisations seek to represent the wider industry, while others confine their operations to a particular sector; and finally, certain organisations function primarily as representative associations, whereas others claim to be 'independent' standard-setting and inspection bodies.

British Organisations

Industry-wide representative associations

The British Security Industry Association

The British Security Industry Association (BSIA) is the largest and highest profile trade association in the UK private security industry. It was founded in 1967 to represent and promote the interests of the industry. BSIA draws its members from most sectors of the industry and claims its member companies are responsible for 70 per cent of the security business undertaken in the UK: the only notable exceptions are door supervisors, locksmiths and private investigators. BSIA is split into eight sections to represent the different sectors of the private security industry (BSIA, 1999a).

BSIA represents the largest security companies and many of the medium-sized firms as well as a few small ones, although membership varies across the different sectors of the private security industry. BSIA is strong in sectors such as transport, with almost 100 per cent of member companies, but weak in others, such as manned guarding, where there are several hundred non-member companies. However, in terms of its members' share of the market, even in this sector BSIA accounts for over 50 per cent of the contract market turnover (BSIA, 1994b). In certain sectors there are many hundreds and perhaps thousands of firms that are not members of BSIA, in addition to those in sectors that the BSIA does not seek to represent. Consequently, a number of smaller firms have objected to BSIA's efforts to portray itself as the voice of the industry, when they regard it as little more than a club for the 'big boys'. However, BSIA is a trade association and, as such, is bound to represent the interests of its members. Even if those members are predominantly larger companies, then it is still compelled to represent their views.

Whatever the debate on how representative BSIA actually is, the association is the most influential body in the industry. It is represented on the boards of the Inspectorate of the Security Industry (ISI) and the National Approval Council for Security Systems (NACOSS) and on numerous European bodies such as the European Committee for Safe Manufacturers Associations (EUROSAFE), the European Committee for Alarm Manufacturers and Installers (EURALARM), the Commission for European Norms Electrical (CENELEC) and the Confederation of European Security Services (COESS). BSIA actively lobbies government, the police, the British Standards Institution (BSI), insurers and other important and influential organisations. The association regularly meets with government ministers and civil servants, and they in turn often speak at BSIA functions. It has numerous committees dealing with the different sectors of the industry, as well as a full-time staff headed by a chief executive and based at its Worcester headquarters (BSIA, 1999a). BSIA once attempted to merge with the International Professional Security Association (IPSA), but in July 1994 its proposals were rejected by IPSA's International Council, in an atmosphere of some acrimony. Subsequently, IPSA's support for the Security Industry Training Organisation (SITO) and ISI was withdrawn (IPSA press notice, 11 July 1994).

The International Professional Security Association

The International Professional Security Association (IPSA) was founded in 1958 as a professional body designed to promote and encourage the science and professional practices of industrial and commercial security. The association was established by security managers disillusioned with their perceived status as 'night-watchmen'. Its aim was to improve the status of members through the exchange of ideas, knowledge, information and experience. Initially, IPSA was a professional association for individual members but it has since grown to include company members and has become both a professional and trade association.

IPSA has members from most sectors of the security industry, including manned guarding, security consultants, security systems installers, security equipment manufacturers and training companies. It tends, however, to represent the smaller and medium-sized companies. IPSA also operates the British Security Registration Board, where individuals can register their curricula vitae, thus allowing a limited form of vetting to take place. IPSA was once regarded as one of the most influential associations, but since the failure of the proposed merger with BSIA in 1994 the association has been in gradual decline.

The Joint Security Industry Council

The problems of finding a body to represent the whole private security industry, and the failure of the BSIA/IPSA merger talks, culminated in the formation of the Joint Security Industry Council (JSIC) in 1995. It was created out of the Security Industry Lead Body (SILB), founded in 1986 to develop consistent training standards and national vocational qualifications (NVQs) for the industry. Funding from the Employment Department ended in 1994 and, as a result, many representatives on SILB decided to use it, instead, as a basis for an industry-wide voice. JSIC's aims are to enhance the influence and identity of the private security industry, and to continue the development of national occupational standards.

JSIC's membership includes a variety of trade and professional associations, inspectorates, trade unions and companies from the whole spectrum of the industry, including virtually every major voice in the industry except BSIA. It remains to be seen whether JSIC will be able to develop a line that is acceptable to its diverse membership, which incorporates many differing views. Given the fragmentation of the industry, with many different and divergent interests, it seems probable that JSIC will be restricted to general statements rather than specific policy initiatives. However, it is too early to determine how successful it will prove.

Other industry-wide associations

There are some other associations that draw a significant membership from other areas of the industry, despite being strongly rooted in the manned guarding sector. The International Institute of Security (IIS), founded in 1968, was once attached to IPSA, but is now independent. Its membership is restricted to those who have passed its Graduate, Member and Fellowship exams. The membership of the Institute of Security Management (ISM) consists primarily of security professionals working in the manned guarding sector, but it also represents other sectors of the industry such as consultancy, investigations and security systems. It was founded in 1988 and now has over 250 members. The American Society for Industrial Security (ASIS) was founded in 1955 as a professional association for security managers in the USA. It now has over 25,000 members in 25 chapters across the world, one of which covers the UK and includes over 400 members from the manned guarding sector, security consultancy, security shredding, security systems and other security sectors. The Security Institute of Ireland (SII) is also relevant here, as some of the SII's Irish members come to the UK to work in the private security industry and retain their membership. Finally, the International Security Management Association (ISMA) is based in the USA and was formed in 1983. At the time of writing, yet another 'Institute' was formed called the 'Security Institute'.

Industry-wide independent inspection/standard-setting bodies

The Security Industry Training Organisation

The Security Industry Training Organisation (SITO) was founded in 1990 to encourage a higher profile for training within the private security industry. It was recognised by the government as the organisation responsible for training and education policy within the UK. Working with representatives from the industry, SITO defines the future direction of training; it then creates training qualifications and markets them to the industry (SITO, 1993a). The impact of some of these qualifications on the services and products of the private security industry will be explored in the following chapters.

Loss Prevention Council

The Loss Prevention Council (LPC) was established in 1986 by the Association of British Insurers (ABIn) and Lloyds as a body to improve the practice of the reduction of losses. It covers aspects of the private security industry, although its terms of reference are much wider. The LPC has within its ambit the Loss Prevention Certification Board (LPCB), which is an independent body that approves products and services in the fire, security and building sectors. Most of the products approved are not security-related, but they cover a wide range of sectors. This is underlined by the LPCB handbook, which lists manufacturers of intruder alarm detection equipment, safes, and security doors and shutters (LPCB, 1999).

Associations representing the manned guarding sector

There is a plethora of associations that represent companies and individuals operating in this sector. BSIA has categories for manned security services and transport, and IPSA has member company, manned guarding and in-house categories for companies and large in-house departments in this sector. There is also an association formed in 1998 specifically for security couriers, the Express Carriers Security Association (ECSA). Individual security managers and security officers can also join IPSA as well as IIS. As discussed above, many managers in this sector may also belong to ASIS, ISM, SII and/or ISMA. There is also a body representing security officers, the Association of British Security Officers, about which very little is known. If trade unions are included in this analysis, the membership of the GMB union includes around 30,000 security officers.

There are also some regionally-based associations that represent security managers and security officers. These include the Essex Security Federation, the Midlands Association of Security Companies and the Southern Counties Chief Security Officers Association. We have also come across a number of other regional associations whose continued existence — and, if they do still exist, the nature of their activities — is impossible to determine. These include: the Chief Security Officers Association (North East); the East Anglian Security Association; the East of Scotland Security Association; the Hertfordshire and District Industrial Security Federation; the West of Scotland Security Association; the West Yorkshire Security Association; and the Western Security Association.

Some close protection officers in the UK are members of the International Bodyguards Association (IBA), which was founded in Paris in 1957 by Major Lucien Ott, with the aims of standardising training, creating a register of bodyguards, representing bodyguards' interests and ultimately bringing responsibility to this sector. IBA now has members and offices in 35 countries. Until 1996 there was no representative association for door supervisors, and then two were formed on the same day — further illustrating the fragmentation of the industry. These were the National Association of Licensed Door Supervisors (NALDS), which has links to the GMB, and the National Association of Registered Door Supervisors (NARDS), which later changed its name to the National Association of Registered Door Staff and Security Personnel (NARDSSP) although it is unclear whether the latter is still in business.

There are also specific associations within the guard dogs sector of the static guarding industry. The British Institute of Dog Trainers (BIDT) is one of the oldest and now represents the wider spectrum of dog trainers. However, but when formed in 1974, BIDT was aimed primarily at security dog trainers, and still has an active security dog section for them. The development of an NVQ for this sector has also led to the formation, in 1996, of the National Association of

Security Dog Users (NASDU), which aims to develop nationally-recognised standards and a code of practice for this sector.

Independent voluntary regulatory bodies in the manned guarding sector

The Inspectorate of the Security Industry

The main voluntary regulatory body specifically designated as covering the manned guarding sector is the Inspectorate of the Security Industry (ISI). ISI was founded in 1992 by both BSIA and IPSA — partly from their existing inspectorates — with the support of the Home Office (ISI, 1998). It was designed to be an independent inspectorate for contract static guarding firms and cash-in-transit (CIT) firms inspecting to BS 7499, BS 7858 and BS 7872 for CIT operators, as well as to the quality standard BS EN ISO 9002. ISI has an independent board and has since been accredited by the National Accreditation Council for Certification Bodies (NACCB) — now the United Kingdom Accreditation Service (UKAS) — as an independent certification body. However, there have been criticisms of its independence from some quarters in the industry; IPSA withdrew from the governing board in 1994 (and has yet to return). Following this decision IPSA re-formed an inspectorate for its members to inspect the sector code of practice, BS 7499. At the time of writing ISI was also about to merge with NACOSS.

Security shredding association

In 1997 the National Association of Information Destruction (NAID) was formed to represent and regulate the information destruction industry. Until then, this growing sector had no representative association or section in a larger association, or standards laid down. NAID has a small but growing membership drawn from the quality companies involved in information destruction, from both static and mobile locations.

Associations representing security consultants

The main voice for security consultants is the Association of Security Consultants (ASC), although it has less than 50 members, this is hardly surprising given the scarcity of independent consultants, who are the only individuals eligible to join ASC's main sections. Founded in 1991, ASC has gained a very high profile during its short existence, despite its limited membership, and is represented on some important boards and forums. Security consultants are also represented by ASIS and IPSA, which have member company categories of membership. Even BSIA represents some consultants in its associate category of membership, although purist observers would question their independence. Burglary insurance surveyors are a type of security consultant, and the body that represents them is the Association of Burglary Insurance Surveyors (ABIS), which was founded in 1953 and has a membership of some 550.

Associations representing private investigators

The UK has a long history of professional associations in this sector. The main organisation representing private investigators is the Association of British Investigators (ABI). ABI was founded in 1970, when it changed its name from the Association of British Detectives. The latter was formed in 1953 from the merger between the British Detectives Association and the Federation of British Detectives, which can be traced back to 1913. ABI has around 500 members and generally represents the reputable private investigators. There is also the Institute of Professional Investigators (IPI), which was founded in 1976 and is open to the 'wider investigative community' such as the police, the military, and forensic investigators. IPI is essentially an academic body offering courses in investigation and it also has a membership of around 500. In fact, many private investigators belong to both ABI and IPI. There are also two other associations based at the same address in Birmingham, and about which very little is known: the Association of British Private Investigators and the Institute of British Detective, Investigative Security and Forensic Specialists. Many private investigators also work as certificated bailiffs, and are represented by the Certificated Bailiffs Association of England and Wales (CBAEW), founded in 1906.

Associations representing installers of electronic security systems

The installation of electronic security systems is the sector most densely populated by associations and inspectorates. A number of alarm and security system installation firms are members of BSIA and IPSA, but there are also a number of other national and regional associations representing installers in this sector. The Electrical Contractors Association (ECA), as its name suggests, primarily represents electrical contractors who design and install the electrical engineering services needed in commercial and domestic premises. However, a significant proportion of ECA's membership are also involved with security systems, and for some it is their main source of work.

There are also a number of other regional trade associations operating within the sector, but representing relatively few workers; information about them is fairly limited. Because of the nature of this sector it is difficult to be precise about the number of trade associations — there may be other small regional bodies who keep a low profile. The National Association of Security Services (NASS) was established in the early 1990s to represent the smaller alarm and equipment companies which were not eligible for NACOSS registration, or indeed did not seek it. In the Wessex area there is a similar association called the Security Alarm Federation (SAFE), based in Swindon and founded around 1994. There is also another organisation with the same name based in Wales, which was also formed around the same time. In the North West, the Merseyside Security Association, founded in 1971, predominantly represents intruder alarm installation companies. We have also come across an organisation called the British Association of Security Installers, although it was not possible to secure any information regarding its operations.

The Mobile Electronics and Safety Federation (MESF), formerly known as the Car Radio Industry Specialist Association (CRISA), was formed in 1978 to represent the installers of electronic equipment in motor vehicles. Originally the majority of firms represented by MESF were almost exclusively concerned with the installation of car radios. However, with the increase in vehicle crime the installation of electronic security systems has become a major part of its members' work. MESF represents its members to government, BSI and the Vehicle Security Installation Board (VSIB).

Independent voluntary regulatory bodies for installers of electronic security systems

National Approval Council for Security Systems

The main voluntary regulatory body in the installation sector for electronic security systems is NACOSS. It was formed in January 1991 as a result of the merger of the National Supervisory Council for Intruder Alarms (NSCIA) and the Security Services Inspectorate (SSI). NSCIA had been founded in 1971 by BSIA. SSI was also established by BSIA, in 1988, as a break-away from NSCIA. NACOSS describes itself as an independent regulatory body with the responsibility for ensuring that the installation of security systems meets the highest standards. It is the only such body in this sector to be accredited by UKAS as an independent organisation competent to certificate, where no single interest predominates. NACOSS is also the largest regulatory body in the sector and dominates the commercial and remote signalling intruder alarm markets.

Security Systems and Alarms Inspection Board

The Security Systems and Alarms Inspection Board (SSAIB) was formed in December 1994, when it took over the inspection role from the Security Services Association. The change meant the creation of an independent board drawn from numerous other interested organisations. It is not recognised as an independent inspection body by UKAS, but is by the Association of Chief Police Officers (ACPO) for the purposes of the ACPO Intruder Alarms policy. The SSAIB has around 200 firms registered with it, mainly in the north of England. It inspects firms to the relevant British Standards, such as BS 4737 for alarm installation.

Other ACPO-recognised intruder alarm installation inspectorates

The Alarms Inspectorate and Security Council (AISC), formerly known as Allied Independent Security Consultants, was founded in June 1990. Under its former name the organisation was engaged in inspections of alarm installers to the relevant British Standards. Following the launching of the new ACPO Alarms policy in May 1995, Allied Independent Security Consultants did not meet the requirements for an independent inspectorate and other criteria. It was therefore faced with the choice of folding or changing into an independent inspectorate: it chose the latter. A number of other associations have either turned into independent inspectorates, or have actually been founded, as a result of the ACPO Intruder Alarms policy. Notable examples include the Greater Manchester Alarms Federation (GMAF) and Integrity 2000, both of which have been recognised as independent inspectorates by ACPO.

Vehicle Security Installation Board

The VSIB was formed in 1994 as the national regulatory and accreditation body for installers of vehicle security equipment. It has the backing of insurers, the motoring public, locksmiths, installers and the Home Office. VSIB has an independent board and inspects those who seek accreditation to the standards of its code of practice. In a sector with an estimated 2,000 to 3,000 installers, VSIB had inspected some 700 firms by the time of writing.

Associations representing physical security equipment installers

Master Locksmiths Association

The Master Locksmiths Association (MLA) was founded in September 1958 as the Greater London Master Locksmiths Association, after a series of meetings in a pub called the Grafton Arms in London W1. The association soon began to grow, and by 1972 the regional mix of the membership warranted the dropping of the 'Greater London' prefix. The association is now both a professional and a trade association representing locksmiths, key cutters and manufacturers in four categories of membership and has over 1,500 members. All members of the association are subject to a strict code of ethics, which covers integrity and honesty, abiding by the rules of the association, impartiality, professional conduct and the promotion of the profession (MLA, 1994). Members are also subject to additional requirements according to which membership section they join.

Associations representing manufacturers and distributors of security equipment

There are a number of trade associations that represent manufacturers and distributors of security equipment. Again, BSIA has a number of membership categories that cover this sector, as does IPSA. There are also a number of other trade associations that represent particular sectors of manufacturing and distribution. The CCTV Manufacturers and Distributors Association (CCTVMDA) was formed in 1994, to represent, as its name makes clear, firms manufacturing and distributing CCTV. Traditionally, firms involved in the installation of CCTV had joined BSIA, but manufacturers and distributors had not joined any association. CCTVMDA has over 25 members and seeks to give a voice to, and gather accurate market research for, this sector of industry (Ta'eed, 1995).

Another body engaged in specialist electronic security equipment is the UK Electronic Article Surveillance Manufacturers Association (EASMA). It was formed in 1992 to promote the use of such equipment by retailers and to raise the standards of its three member firms. To achieve this, EASMA has a code of conduct that all members must subscribe to, and produces promotional literature advocating the equipment (EASMA, u.d.). Those firms that manufacture electronic vehicle security products are represented by the Electronic Vehicle Trade Section of the Society of Motor Manufacturers and Traders.

There are several associations engaged with physical security products. The Guild of Architectural Ironmongers, founded in 1961, primarily represents distributors and manufacturers of ironmongery. Many of these products are security-related, such as access control systems and locks and bolts — essentially, any product fitted to doors and windows. Most lock manufacturers are members of the British Lock Manufacturers Association (BLMA). BLMA represents all the major firms in this sector, which together employ some 5,000 individuals. Shutters and doors are also clearly physical security products, and manufacturers are often members of the Door and Shutter Manufacturers Association (DSMA). It has 26 full members and 30 affiliates, with a combined turnover of at least £100 million, and estimates there are 100 firms outside its membership. In the laminated glass sector (safety and security) most British manufacturers are members of the Glass and Glazing Federation. Many fences are installed for purely security roles, and the Fencing Contractors Association represents companies involved in this sector. There is also a training body, the National Fencing Training Authority. The Security Seal Industry Association represents the manufacturers of security seals.

The increasing spread of the private security industry into areas that were formerly the exclusive preserve of the police and prison service has meant that security firms have had to purchase specialist equipment such as hand-cuffs, riot control equipment and specialist uniforms. These products are generally only available from specialist firms; those that provide such products are often members of the Association of Police and Public Security Suppliers (APPSS), which is a division of the Defence Manufacturers Association.

Associations representing users of security equipment and services

There are very few associations that solely represent a specific group of security service or equipment users. In 1996, however, the CCTV Users Group was formed providing a forum for users of CCTV, including local authorities, hospitals, the police and schools, to share their experiences and views of the technology. One of the main functions has been to develop a code of practice for the use of CCTV.

Organisations representing security in specific areas of commerce, industry and the public sector

Certain areas of commerce, industry and the public sector have also developed their own security association. In hospitals and other related institutions the National Association for Healthcare Security (NAHS) has been created. It represents those responsible to the National Health Service (NHS) Executive Policy Security Group for security in the NHS. Security managers working for the top hotels also have their own security association, the Institute of Hotel Security Management (IHSM). University security managers have the Association of University Chief Security Officers (AUCSO), which has almost 150 members. Many security managers involved in the defence industry are members of the Guild of Security Controllers (GSC), founded in 1963 as a professional association for security managers at MoD 'List X' companies (see chapter 15), and as a forum for the exchange of information. GSC now has over 350 members, whom it also represents at the quarterly meeting of the MoD's Liaison Group for Defence Industrial Security (LGDIS) as well as in JSIC. The oil industry has the British Oil Industry Security Association, and in the food manufacturing sector there is a body entitled the Food Distribution Security Association. In Scotland there is an association for those with responsibility for security in museums, art galleries, historic houses and other similar establishments, called the Scottish Association for Museums Security. Co-operative associations also have a security association, the Co-operative Security Services Association. For those employed as security managers in breweries and allied trades, there is an association called the Institute of Brewery and Allied Trades Security Managers.

Other relevant associations

There are a number of associations that are dedicated to the private security industry but which do not fit easily into any of the above categories. One of those associations is called Ex-Police in Industry and Commerce (EPIC), formed in 1979. The association, which now has over 400 members, represents retired police officers working in security in commerce and industry. Another association, which probably has the most exclusive membership in the industry, is the Risk and Security Management Forum (RSMF). RSMF was founded in 1990 and comprises the most senior individuals in the industry; membership is by invitation only. Essentially, RSMF seeks to further the aims of the business concept of risk and security management. This objective is pursued by holding seminars with high-profile speakers.

Informal associations

The sensitive nature of some aspects of the security industry, and the common background of certain individuals working within it, has led to a number of informal security groupings. Often these have no constitution, name or formal arrangements. Instead, they are often only informal gatherings of security personnel at a lunch or in a public house on an ad hoc basis. Obviously these informal associations are difficult to identify, and there are many that we have probably not discovered. Some of the associations that we are aware of include an informal group of ex-Special Branch officers now working in the security industry, who meet for lunches to discuss issues of mutual interest. There is also a group of senior security managers working for merchant banks, who regularly have informal meetings. Another high-ranking group of security professionals meet under the auspices of the Sellwyn House Group. In the leisure security sector of the industry there is no professional or trade association for those firms involved in providing security for special events; however, most of the main companies meet on an annual basis for a conference to discuss issues of relevance. These conferences have also been instrumental in establishing training standards for that sector. Such meetings could be described as an example of an informal association.

International Organisations

At an international level there is also a range of organisations that have emerged for the private security industry. They can be divided into those consisting of a confederation of national security associations, and those that represent security firms and individuals across international borders. Some have already been mentioned, as they have a significant membership in the UK. There are also some associations which claim to be international but have only a few members outside the host country; these have been excluded from our analysis. The focus of our research centres upon Europe and North America, but there may be significant international organisations in other parts of the world. There are also some international bodies that are not security-related but which have security committees. For instance, in the field of aviation security both the International Air Transport Association (IATA) and the European Civil Aviation Conference (ECAC) have large interests in security. However, as they are not primarily security organisations, they have been excluded from this analysis.

International confederations

COESS was established in 1989 to include the main manned security trade associations in Western Europe. The major aim of COESS is to participate in the work of harmonising EU countries' national legislation in the manned guarding sector. To achieve this, COESS undertakes and publishes research, defines the common interests of its membership and represents them to European organisations. EUROSAFE, which was founded in 1988, undertakes similar functions to COESS, although it has a more important role in representing safe manufacturers on the Commission for European Norms (CEN), the European Standards body. EURALARM is a similar association for alarm manufacturers and installers and was founded in 1970. It also represents its members on both European standards bodies — CEN and CENELEC. Another important European federation is the European Locksmiths Federation (ELF). A federation of European certification bodies, the European Fire and Security Group (EFSG), allows for the certification of one product by a single body in order to gain acceptance by all the certification bodies of its member states. It covers the product areas of fire alarms, gas and water extinguishing systems, intruder alarms and safes.

The international representative association for private investigators is the Internationale Kommission der Detectiv-Verbände (International Federation of Associations of Private Detectives), or IKD as it is commonly known. It operates through its national member associations and represents over 50,000 individuals worldwide. It has even gone as far as developing a draft directive for the European Commission, to introduce minimum standards throughout Europe for private investigators — although there seems little interest from the Commission in this area so far.

International associations

Perhaps the oldest and most prestigious international association is the Ligue Internationale de Sociétés de Surveillance (LISS), which was founded in 1934. The Ligue is an international association that represents the leading national security organisations, and will only accept the leading companies from each country as members. It aims to broaden contacts through exchanges of experience and opinion, in addition to deepening the understanding of its members and the countries they represent. The main forum for this is the Ligue's bi-annual General Assembly. The Ligue has members from countries ranging from Belgium to Tunisia.

In 1976 the European Security Transport Association (ESTA) was formed to represent cash-in-transit companies in Europe. Its aims are to undertake and publish research and represent members to European organisations. Another representative association for manufacturers of security equipment is the European Association of Security Equipment Manufacturers (EASEM), which was founded in 1987. Its formation was partly a response to the perception that German manufacturers were dominating EURALARM. The aims and functions of EASEM are typical of all European associations.

There is also an association that is dedicated to the identification of computer security problems and to the development of solutions for them. It is called the European Security Forum (ESF), and was founded in 1989. ESF's members include many large companies from throughout Europe. It is not, however, a representative association, which makes it unique among national and European organisations.

Conclusion

We are aware of over 60 associations, inspectorates and standard-setting bodies in the UK. An exact number cannot be given with confidence because of the difficulty in finding out if some associations still exist; others simply may not have been identified. Our estimate should be taken as the best indicator of the total number of private security-related organisations in the UK.

Much of the growth in the number of organisations has reflected the continuing fragmentation of the industry, with new bodies emerging to represent and set standards for sectors of the industry as each develops and matures — ASC and VSIB are examples of such a process. New associations have also emerged because others were not adequately representing their members' interests. JSIC, for example, was formed to provide a more representative voice for the industry. In other sectors, the demands of quasi-regulators, such as the police and their 1995 ACPO Intruder Alarms policy, have intensified fragmentation. Other bodies, such as AUSCO and IHSM, have emerged as forums for security personnel in particular areas of commerce, industry and the public sector, rather than developing through existing security associations. Finally, some bodies, such as ISM, have emerged as a result of a split in an existing association.

Despite the increasing fragmentation of the industry, attempts at consolidation and merger have generally been resisted. The failure in 1994 of the merger (or takeover, as some critics claimed) of the two largest associations, BSIA and IPSA, is illustrative of this. Perhaps the only exception was the formation of NACOSS from the NSCIA and SSI, though SSI had itself broken away from the NSCIA in the first place. The merger of ISI and NACOSS, however, may mark the beginning of the end of fragmentation, as the potential consequences of regulation lead many organisations to consider merging, so as to be in a better position to deal with the new regulatory framework when it emerges.

The number of international associations is also increasing, particularly Europe-wide organisations. This can be largely attributed to the pressure to meet the challenges of European standard-setting, and to the increasing number of decisions of importance being taken at a European level. Given this increase, there might be a further increase in the need for Europe-wide confederations and associations. At the same time, this increasing importance may expose the divergent interests of the national associations of different countries, and thus fuel splits in European confederations and associations.

Chapter 6

Manned Security Services

Most people envisage the private security industry in terms of uniformed personnel — what is commonly called the manned security services sector. This sector encompasses more than uniformed personnel, however. 'Manned security services' is a generic term used to cover organisations offering (and using, in the case of in-house security staff) trained personnel to provide directly a range of security services. The term covers services that provide a security function, and it is the actual provision of this function that distinguishes the sector from others in the industry. An alarm installer provides a service in installing a security product, but it is the alarm which provides the security function — whereas security guards patrolling a building are themselves fulfilling a security function. The definition as it stands might also include private investigators, security consultants and private-sector detention services. Nevertheless, a distinction can be made between the 'professional characteristics' of the first two and the distinct services provided by the latter.

Manned security services can be further divided into sub-sectors according to their structure and the activities they undertake. The following sub-sectors have been distinguished: static manned guarding services (contract and in-house); cash-in-transit (CIT) security services; door supervision and stewarding services; and close protection services. Some of the characteristics of and differences between these sub-sectors will be explored later in this chapter; each is becoming increasingly distinct as the process of fragmentation in the industry continues. Where 30 to 40 years ago firms would often have provided services in all these areas of the industry, they are now increasingly concentrating on one specific such area.

Static manned guarding services

Like that of so many activities in the private security industry, the definition of static manned guarding services presents certain difficulties. It is difficult to find a coherent and all-embracing definition, which captures the essence of the sector. BSIA (1994a:1) has defined it as 'static or mobile security services used in the guarding of publicly and privately owned premises'. Many observers, however, regard this definition as unsatisfactory, because it excludes CIT services, prison and escort-related guarding and 'bouncers' (or door supervisors, as they are now more commonly known). However, the inclusion of the word 'mobile' in this definition is confusing, as it implies movement and would therefore suggest (rather than exclude) CIT services. BSIA's definition is typical of most previous efforts, in failing to articulate clearly what manned guarding is. We believe it is possible, despite the problems, to construct a workable definition of static manned security services, as follows:

> The provision of predominantly static security services — whether by an in-house organisation, a contract company or a combination of both — based on guarding, patrolling, searching, surveillance and/or the enforcement of laws and organisational rules by predominantly full-time and uniformed personnel.

Our definition was arrived at by analysing the activities of a number of security firms, which claim to offer static manned guarding services, and by extracting — common functions: guarding, patrolling, searching, surveillance and the enforcement of rules and the law. Some of the services undertaken in this sector cross the boundaries of the core activities identified above, which is inevitable given the wide range of activities, in often multifaceted situations. Access control provides a perfect example: it serves a guarding function, keeping out those who are not supposed to be in a particular area, as well as an enforcement function, enforcing organisational rules as to who may or may not enter that area.

Guarding is one of the most common activities of security officers working in this sector. Some of the functions it entails include access control, alarm response, protecting life and property, key-holding and patrolling. The duties of most security officers include a guarding role. In many respects patrolling is a function of guarding, but it is such a common activity of static guarding firms that it deserves consideration in its own right. Security officers are employed by a range of organisations to patrol internally and externally in order to deter crime and look for potential problems and hazards.

Searching and screening duties are another core function for many of those working in this sector, often in organisations vulnerable to terrorist attack. Such procedures cover people and vehicles entering a building or other location. The most common occurrence is at airports, where passengers and luggage are often screened by x-ray machines for explosives, weapons, etc. Surveillance functions are among the most important duties of many security personnel in this sector. The huge increase in the use of CCTV has resulted in a growing number of security officers employed in monitoring CCTV, and hence in a predominantly surveillance function. Surveillance can also be accomplished without the benefit of electronic equipment; other methods include patrolling, and using mirrors and listening devices. Security officers may be engaged in the monitoring of alarms for security, fire or purely commercial purposes, for example the temperature alarms of freezers. The monitoring may be conducted in a central station, control room or even a gate house. The enforcement of rules and the law are also common roles for security officers, although most do not have any special statutory powers (most can derive powers from other sources — Sarre, 1994). Most security staff will be involved in the enforcement of some organisational rules, and many may also enforce public laws.

Nevertheless, our definition of static manned guarding services also identified a number of other key characteristics of this sector. The first was the 'provision of predominantly static security services'. This does not literally mean that all the services must be static. The phrase was used to exclude predominantly transport-orientated security services such as the CIT sector. Some transport-based security services, such as mobile patrols and secure escorts, have been included within the 'static' guarding sector, as we regard the majority of their activities as more appropriate to this sector than to CIT. Another key factor in our definition of static manned guarding services was the use of 'predominantly full-time and uniformed personnel'. This reflects the nature of a sector where the vast majority of security officers are full-time, and wear a uniform (which is often quasi-police or military in style). Static manned security services are employed in a variety of organisations and undertake a wide array of functions. We have concentrated on the main responsibilities of security officers in this sub-sector. Some of their other duties are not listed due to space considerations and to the fact that these other duties are not security-related. We would also include within the scope of this sector private wheel-clampers, as there are many security guards, and some security firms, who undertake this role as part of their wider duties and they would also fit in this sector by virtue of 'enforcing organisational rules'. It is also clear from the core functions discussed above how those employed in this sector generally undertake a combination of crime prevention, loss prevention, order maintenance and protective functions, which indicates that the sector is clearly part of the private security industry.

Size and structure

Static manned security services can be divided into contract and in-house sectors — although some organisations use both. The in-house sector can be defined as:

> An individual or group of individuals directly employed by an organisation, for the primary purpose of providing manned security services for that organisation (Button and George, 1994: 210).

It is often referred to as the proprietary sector, particularly in North America. The in-house sector is where an organisation directly employs security officers to undertake static guarding functions. The contract sector, however, is composed of firms that offer static guarding services for hire. Hence when an organisation employs a contract security firm to carry out static guarding functions, the security officers used are employed by the contract firm, rather than by the organisation they protect. This distinction is sometimes disguised, as many contract security officers are now asked to wear the uniform of the firm they are guarding.

Static manned security services accounted for a turnover of just under £1.5 billion in 1998, a decline of 10 per cent from the estimated size of the industry in 1993 (BSIA, 1994a; 1999b). The same research estimated there were approximately 125,000 security officers in 1998, a decline from the 1993 estimate of 130,000. There have been other attempts to gauge the size of this sector, most notably by the Policy Studies Institute (PSI), which stated there were 129,257 security officers in 1994 (Jones and Newburn, 1994), suggesting a realistic figure of perhaps 125,000.

The market ranges from the multinational companies such as Group 4, Rentokil Initial and Securicor, which have total turnovers of hundreds of millions of pounds and employ many thousands of security officers, down to the 'one man and his dog outfit' (McManus, 1995). This market also includes companies involved in many other areas of the industry, such as Securicor, which operates in CIT, alarms, private prisons and courier and custody services. It also encompasses companies solely involved in static manned guarding. By contrast, some companies that operate in the contract guarding sector also have large interests in other industries; Rentokil Initial, for example, is also involved in pest control, cleaning, medical work, tropical plants and facilities management services.

The market structure of the private security industry drives down prices and standards for a number of reasons, of which one of the most important is the ease of entry into contract guarding. All that is required to establish a static guarding firm is a second-hand fireman's uniform, headed note paper and a dog; there are no specific legal requirements for entering the market. Therefore, somebody could leave prison, set up a security firm with virtually no capital, start bidding for contracts and actually win them — and this does happen. The attitudes of many purchasers of security services, for whom price is often the only consideration, have not encouraged high standards.

Some organisations are more rigorous in their selection of security companies and review a range of criteria, such as membership of trade associations, adherence to industry standards and quality of training; but such organisations are regrettably in a minority. In a survey for BSIA of those who buy security services, 24 per cent (even after being prompted) had not heard of BSIA, and 49 per cent were unaware of IPSA's existence (BSIA, 1994c). If those buying security services have not even heard of the two main trade associations, there seems little chance for companies with higher standards to win more contracts. This is not to say that all those operating in this sector who are not members of BSIA or IPSA are companies with low standards; neither would it be accurate to suggest that all members of the two associations automatically provide a good service. Nevertheless, members of the two main associations must have met certain minimum standards and this is not the case with the majority of non-members. Ultimately, much of the contract market is driven by price, as the undertaking offering the cheapest deal wins the contract, irrespective of its standards. Unfortunately, some major organisations purchasing security also fall into this trap — including government departments (see House of Commons Defence Committee, 1990). The consequence of this market structure is a downward spiral in costs and quality, as firms enter the 'Dutch auction' of cutting costs to survive in the market.

There is also a sizeable in-house segment of the static guarding sector. Indeed, some in-house security forces are as big as some of the larger contract companies; for example, the MoD Guard Service employs around 4,500 officers. The government has its own security force, called the Security Facilities Division (previously known as the Security and Facilities Executive — SAFE), which consistis of some 700 guards, as well as a number of other in-house forces. The Metropolitan Police also employs almost 400 in-house civilian security officers. The in-house sector is different from the contract sector in a number of respects. Research has illustrated that standards are generally higher than the industry minimum standards (Button and George, 1994). For a more detailed analysis of the in-house sector, including why some organisations prefer it, see Button and George (1994 and 1998).

In the last few years a structure of training has emerged for the static guarding sector that barely existed before the formation of the Security Industry Training Organisation (SITO) in 1990. Prior to SITO's formation, qualifications for security officers were restricted to BSIA's basic two-day induction course and IIS's Graduate, Membership and Fellowship qualifications. The BSIA course was very basic, while the IIS qualifications were aimed more at managers, and excluded many ordinary security officers from applying on ability and cost grounds. Since SITO was formed, a range of generic and specialist training qualifications have been developed for this sector. However, the structure of the sector has meant that only limited numbers have sought to acquire these qualifications. Firms often fear that investment in training will raise costs and damage their chance of winning contracts, and that the guards, once trained, will move on to another company which offers a higher rate of pay. These fears are well-founded, with estimates of annual labour turnover in the contract sector ranging from 30 to 40 per cent. For managers and those aspiring to such positions, there is in addition to the IIS qualifications — the Certified Protection Professional (CPP) qualification of ASIS. In the last ten years there has also been a growth in undergraduate and post-graduate qualifications in security management.

The guard dogs sector

Dogs (and other animals, such as geese) have been used since the very beginning of modern society to guard property. It is therefore not surprising to find firms that offer guard dogs and handlers for hire, and there are also firms dedicated solely to training dogs and/or handlers. This constitutes a distinct part of the sub-sector, the static guarding sector.

There are two types of dogs generally used by security firms. 'Generalists' are trained to bark and look vicious under certain circumstances, to patrol an area with a handler and/or to guard an area on their own. Second, there are 'specialists' dogs that are trained to search for specific substances such as drugs and explosives. The vast majority of security dogs are 'generalists', although there is a demand for specialist dogs in distinct areas such as aviation security. While many firms offer specialist dogs, we are aware of only one private firm that is licensed by the Home Office to hold explosives and drugs for their training. Those firms that offer specialist dogs without access to these substances must have either unique training methods or illegal access. The number of dogs and dog security firms is difficult to assess. At a recent national conference for such firms it was estimated that there were in the region of 75 firms, together supplying around 1,000 dogs (National Association of Security Dog Users Annual Conference, 1996). Most at the same conference agreed that these figures were probably an underestimate. The figures also exclude the many security officers who, it was claimed, use their pet dogs as guard dogs.

The guard dogs sector is as easy to enter as static guarding: there are no legal or training requirements specific to guard dogs. There are requirements regarding the welfare of dogs and also the upkeep of belonging to kennels those who own a certain number of dogs, or who hire out their facilities for other dogs to stay at, but these are not security-related. There is also the *Guard Dogs Act* 1975, which will be discussed later; however, this legislation only relates to how guard dogs are used. There are not even any nationally recognised training standards for security dog trainers, dog handlers or the dogs themselves, although the recently formed National Association of Security Dog Users (NASDU), is developing them. A dog with the wrong qualities being controlled by handlers with little or no training could be as dangerous as giving them a gun. The risk of serious accident would appear to be considerable.

Regulatory structure

Though the static manned guarding sector as a whole is not subject to statutory licensing, there are specific areas of the sector that are subject to some statutory controls. Before these areas of the industry are discussed, we will examine the voluntary regulatory structure.

Voluntary regulation in this sector is based on the optional membership of trade associations and on registration by an inspectorate. The two trade associations are BSIA and IPSA. All members of the BSIA manned guarding section must register with ISI, or with a certification body that is recognised by UKAS. Non-members of BSIA are also invited to register with ISI. The static guarding sector registration with ISI is based on various forms of membership, which are themselves based on adherence to quality assurance standards (BS EN ISO 9002) and/or relevant British Standards. The latter include BS 7499, the code of practice for manned security services, and BS 7858, the standard for vetting personnel employed in a secure environment. ISI also regulates the cash-in-transit (CIT) sector, which will be discussed later in this chapter (ISI, 1998), and at the time of writing BS 7958 (Management and Operation of Closed Circuit Television Monitoring) and BS 7960 (Door Supervisors and Stewards) had been published, on which ISI intends to offer assessment services (ISI press information, 19 November 1999).

However good this regulatory framework may appear, in the final analysis it is voluntary. There are over 1,000 firms that are not members of BSIA or IPSA and the vast majority of these will make no attempt to meet BS 7499. Nearly all the firms in this sector, including BSIA and IPSA members, have no access to criminal records and thus rely for vetting on work history checks and other methods. This will soon change, once the provisions of the *Police Act* 1997 are implemented and licensing of all security employees and managers introduced. Until then there will remain fundamental weaknesses in the current methods of vetting security staff.

Some of the legislation impacting on the static guarding sector can be divided into statutory licensing and regulatory measures. The one licensing statute affecting the industry is the *Northern Ireland (Emergency Provisions)* Act 1996 (which is renewed every five years) — the unique political situation in Northern Ireland having led to the introduction of controls to license private security services. The *Guard Dogs Act* 1975 was enacted to regulate the keeping and use of guard dogs. It sets out conditions for the control of dogs and guard dog kennel licences. Only sections 1, 5, 7 and 8 have been implemented, by Statutory Instrument (SI) 1767/1975. Sections 2, 3, 4 and 6 have still to be introduced, and these are the most important parts as they set out the licensing requirements. The only section that applies to the industry is the one setting out the conditions controlling guard dogs.

Aviation and maritime security are regulated comprehensively by the Department of the Environment, Transport and Regions (DETR). The legislation affecting these sectors consists of the *Aviation Security Act* 1982 and the *Aviation and Maritime Security Act* 1990, amongst others. The legislation regulating aviation security will be considered in more detail in chapter 14. To summarise, the myriad of legislation, amongst other functions, sets out a framework for minimum security standards on such issues as vetting (although access to police records has only recently been granted), training of staff, screening and searching. These arrangements are inspected for compliance by DETR inspectors. The legislation also gives designated security officers powers of search. Similar provisions are set out for the Channel Tunnel under the *Channel Tunnel (Security) Order* 1994 (SI 570/1994, issued under the *Channel Tunnel Act* 1987). The provision for the railways was issued under the *Railways Act* 1993. Provisions for the contracting-out of court security were incorporated in the *Criminal Justice Act* 1991. This legislation will enable court security officers to search individuals and to exclude or remove them from court. The *Criminal Justice Act* also makes it an offence to assault or obstruct a court security officer.

Cash-in-transit security services

It is not unusual when passing through a town centre to see an armoured car parked outside a bank; indeed, they often seem more numerous than police vehicles. The high public profile of this sector is reinforced by the occasional armed robbery, where staff are subjected to acts of violence and sometimes even worse offences. CIT services have a number of superficial similarities with the static guarding sector. A deeper analysis, however, reveals some fundamental differences in the structure and characteristics of this sector and in the problems associated with it.

The main activity performed by the CIT sector is the transportation of valuables. The main clients of the security companies are banks and large retailers, which use them to transport large sums of cash and other valuables, such as gold. Banks are also increasingly using CIT firms to service their Automatic Teller Machines (ATMs). CIT firms are also expanding their services to smaller retailers — the establishment of the National Lottery has meant many small retailers have large sums of cash on their premises. The sector also encompasses the (declining) activity of wage packeting, in which wages are made up at a secure centre and then delivered to the client for distribution to employees. CIT firms also deliver large sums of cash to clients who make up their own wages.

At this juncture it is important to distinguish the CIT sector from companies offering courier services. Superficially they may seem to offer similar services, with the carriage of a client's property between locations. The difference, however, is that CIT firms always carry valuables.

Couriers transport documents, letters and packages, the majority of whose contents are not valuables and do not need special security protection. The tiny minority that perform security-related functions warrants the inclusion of couriers within the margins of the private security industry, but only as one of the marginal security activities (which will be briefly discussed in chapter 11). Based on these observations we can offer a definition of the CIT sector:

> The provision of secure transportation services for valuables such as cash, coins and precious metals by predominantly full-time uniformed personnel in specially adapted vehicles.

The CIT sector's functions include protection, crime prevention and loss prevention roles, thus clearly placing it within the bounds of the private security industry, although the function of order maintenance would not apply.

The CIT sector has been estimated to be worth around £382 million (BSIA, 1999a). The numbers employed in the sector are subject to divergent estimates; they range from 8,000 (Allen 1991), to the 25,000 estimated by ESTA (personal communication). ISI, which effectively regulates all the companies operating in this sector, estimates that there are over 20,000 employed. However, senior managers from Securicor estimated the figure was nearer 15,000. Given these differences the best guess would be to err on the side of caution with the latter figure.

The structure of the market is very different to that of the contract guarding sector. First, there are many barriers to entering the market. To start a CIT company a large capital outlay is required for purchasing the armoured cars, secure depots, vaults and other specialist equipment. An armoured car can cost as much as £60,000, and while other transport companies may be able to sell their older vehicles to recoup some of their capital outlay, there is no such market for second-hand armoured cars. Furthermore, any security company that seeks to carry the valuables of an organisation needs not only its clients' trust but also insurance cover, which is not easy to gain.

If a CIT business wishes to remain profitable it must also have a large network of clients, or a good 'route density', as it is often described. It is not enough for a CIT firm to have a large number of clients — they must also be in close geographical proximity if the business is to succeed. For example, if one CIT firm has 100 pick-ups spread around 50 town centres the vehicle will have to make 50 trips, whereas a firm with the same number of pick-ups in only 10 town centres will have to make only 10. These networks take many years to develop, and the importance of route density may also explain why some small regional CIT companies have survived in this cut-throat market — Security Plus Ltd in the Midlands is a thriving example.

The barriers to entry would lead one to expect that there would be few suppliers, operating almost as an oligopoly and gaining large profit margins. There are indeed few companies operating in this sector, with the top four firms accounting for over 90 per cent of the market. The largest companies include Securicor, Securitas, Royal Mail Cashco, Brinks, Group 4 CIT Scotland and Security Plus.

There is intense competition in the CIT sector in the UK and there are a number of reasons for this. First, there are very few purchasers of CIT services. Most of the market's services are bought by the four main banks in England and Wales, the two main banks in Scotland and the main retailers; the rest of the market accounts for a very small proportion. These organisations, particularly the retailers, have centralised purchasing systems, which are among the most efficient in the UK. Second, all the main CIT firms have almost identical standards, as members of BSIA and inspected by ISI, so the main basis for distinguishing between them is the price they charge.

Thus a bank or retailer will in most instances change from one CIT firm to another simply if offered a lower-priced contract, as it is likely to receive the same standard of service. In the UK market situation, if a CIT firm loses one contract with a bank, it could be losing up to a quarter of its business. This leads to the overriding aim of CIT firms to hold on to their contracts at any cost. The result is intense competition based almost entirely upon price. This situation has been exacerbated in recent years by the entry of the Royal Mail's Cashco into the market.

Regulatory structure

There is no statutory framework regulating this sector although, as with the static manned sector, there is a voluntary framework. It is also one of the few examples of a successful voluntary framework in the private security industry, as virtually every company operating in the sector meets the relevant standards; almost all are members of BSIA, which means they must register with ISI or another UKAS-approved inspectorate. Pressure from insurers also acts to force most of these companies to meet standards and to join the relevant association and inspectorate. To become registered with ISI, a firm must meet BS 7872, the *Code of Practice for Operation of CIT Services (Collection and Delivery)*. This covers, first, requirements on company organisation, such as the submission of audited accounts, sufficient insurance cover and a central administrative office; second, the code of practice requires the company to demonstrate a proper vetting of staff and to possess adequately protected vehicles; third, the company should meet the relevant training standards; fourth, it must be able to demonstrate proper operations procedures; and finally, it should have proper documentation. Some CIT companies also operate in other segments of the industry, which allows some staff to work in other sectors for overtime. There is also BS 7931, *Code of Practice for the Secure Carriage of Parcels*, which is also relevant to some CIT firms.

The door supervision and stewarding sector

Door supervisors (or 'bouncers' as they are more commonly known) and stewards are often regarded by the mainstream of the private security industry as a peripheral sector, and possibly not part of the industry at all. A deeper analysis of their activities, however, reveals that they have clear security roles. As will be shown, they are all engaged to varying degrees in order maintenance, crime prevention, loss prevention and protection. Meanwhile, there is a clear distinction between door supervisors and stewards: the former work at pubs and nightclubs, whereas the latter operate at pop concerts, sports events and other public functions (and there is a further distinction between stewards at pop concerts and those at sports events).

Despite these necessary qualifications there are characteristics that bind these distinct occupations together. Door supervisors and stewards both perform security roles; they both tend to work in the entertainment/leisure sector; and the majority in both occupations are part-time. In many ways the roles they perform could be described as those of static guarding in the entertainment industry. Hence the definition we offer of the door supervision and stewarding sector contains similarities to that of the static guarding sector:

> The provision of security and crowd-management/control services — whether by an in-house organisation, contract company or combination of both — predominantly for the entertainment sector, based upon guarding, patrolling, searching, surveillance and the enforcement of laws and organisational rules by predominantly part-time personnel.

The five core activities identified earlier in this chapter for the static guarding sector are also relevant for door supervisors and stewards. The primary role of a door supervisor is monitoring admittance to the night-club or pub in accordance with organisational and legislative requirements. There is an access control role, which combines both guarding and enforcement roles. The door supervisors guard the organisation's property by ensuring 'troublemakers' do not enter. They also ensure compliance with company policy, for example by only admitting people who are smartly dressed, and enforce the law by excluding those under age. The access control function also involves a surveillance role, in assessing those entering the premises to see if they are drunk, and may also involve searching customers for drugs and weapons. Inside the club or pub the role of a door supervisor is to identify and prevent incidents such as fights, drug-taking, other crimes and emergencies. This role involves patrolling the interior and keeping areas under surveillance, as well as enforcing organisational rules and the law. Troublemakers may have to be ejected or even arrested by door supervisors, using minimum force. Additionally, door supervisors will have roles in the event of an emergency, as well as other functions such as guarding the cash room and escorting those carrying cash.

Stewards at sports events such as a football match also assume many responsibilities, which fall within the above categories. A typical steward will have an access control role in operating turnstiles. This may also involve organisational rule and law enforcement roles if he or she is charged with looking out for those who are banned from the ground. Other stewards may be responsible for searching fans entering the ground/venue, and those in the ground are often responsible not only for ushering fans to their seats but also for enforcing club rules and laws. They may have to stop fans shouting racist abuse or running on to the pitch, and are empowered where necessary to eject troublemakers. Some stewards may also guard emergency entrances and exclusive areas such as the Directors' box. Hence, there are guarding, surveillance and organisational rule and law enforcement roles. In most grounds there is usually CCTV and this too is often monitored by a steward. The surveillance role is universal, as all stewards will be monitoring the crowd for signs of distress or misbehaviour. They will also have specific roles in the event of an emergency, hence a protection of life and property (guarding) role.

Not all sporting events have stewards performing such functions. In some venues the steward fulfils little more than the 'old image' of an old age pensioner who tears the tickets at the entrance and then watches the match. This is becoming a rarity, particularly in football, as stewards have assumed increasing responsibilities and may even replace the police. Some games played in the Premiership, for instance, have no police inside the ground, and in such situations stewards have an even greater security role. Similarly, in other sports such as boxing, cricket and rugby stewards may soon assume similar roles.

The arrangements for stewarding at a pop concert are slightly different. In addition to the responsibilities already discussed, a steward at a concert has a number of specialised roles, which include guarding functions such as access control in the backstage area. Organisational rule and law enforcement functions include patrolling the crowd to maintain order and to prevent and detect crime. However, the most striking difference between a steward at a concert and one at a sporting event is the role of the former in the stage area, which has become more a crowd management function. This implies the proactive planning of security and involves an understanding of crowd dynamics, flow patterns and the nature of the performance, with the aim of minimising the risk of problems such as people being crushed (Upton, 1995).

Size and structure

Analysing the size and structure of this sector is made difficult by the paucity of research. Only stewarding at football grounds has been the subject of any form of rudimentary research, but this has invariably been into football hooliganism. Apart from this limited research and the occasional short article there is very little literature to build upon. Thus many of the estimates that follow are based on the testimony of experts rather than sound research.

There are differences between the structure of the door supervision and stewarding sectors. Door supervisors are generally employed by clubs and some pubs. They are either employed in-house or by a specialist contract company, or by a combination of the two; there is no research, however, to suggest the relative balance. According to SITO there are approximately 10,000 door supervisors (*The Guardian*, 15 August 1994), but the GMB estimate is around 50,000 (*The Guardian*, 15 April 1996) — although neither source offers any empirical research. However, given that in the City of Westminster alone there are around 1,500 registered door supervisors, the higher figure seems more realistic. Despite such estimates of the numbers employed, there are none regarding the market value of the sector. Given that labour accounts for over 90 per cent of the costs in this sector, an estimate can be made based on the average pay of door supervisors. As many are part-time, a figure of £50 per week for average earnings would seem reasonable — although probably something of an underestimate. If that figure is multiplied by 50,000 and by 52 weeks and rounded-up to allow for other costs (such as full-time door supervisors), the market would be worth an estimated £150 million.

There are more barriers when trying to estimate the number of stewards, because the number of events that they work at is difficult to assess. There are stewards employed at football matches, at other sporting events and at pop concerts in stadiums and indoor venues. At one football match alone there could be over 500 stewards employed, as at Old Trafford (Manchester United), and over 1,500 stewards were present at the 1996 'Who' concert in Hyde Park. To arrive at a very rough estimate of the number of stewards, one could multiply the number of football league clubs (92) by a mean number of 100 stewards per club, to give 9,200. If the stewards at other events were then added, a total figure of 12,000 would not seem unreasonable. In terms of the value of this sector, research for the Football Trust Survey estimated that in the 1994–95 season over £8 million was spent on stewarding by football clubs in the main five English divisions (*The Observer*, 11 January 1998). When the other sectors that use stewards are considered, a reasonable estimate would probably be £12 million for this sector in 1995.

There are hundreds if not thousands of firms involved in the door supervision sector. These are generally small, and there is no dominant national company. Many door supervisors are also employed in-house, as are stewards in football clubs, arenas, universities, etc. There are some large contract security firms that provide stewards for events such as Wimbledon and golf tournaments; however, their interest is more with the broader special events sector, where security is required for a 'one-off' event such as a conference, exhibition, show or VIP visit. More commonly, providers of stewards to sporting events are security firms such as Showsec International, which specialises in crowd management and is a member of the small group of companies which undertake almost all the stewarding of pop concerts at the larger venues. This is the only part of the sector with a coherent structure.

The reasons for the more coherent structure of pop concert security are the greater expertise required of the firms and the very 'closed' nature of the music business community. All the firms and promoters involved know one another and are generally hostile to those companies that try to enter the market with no previous links or experience. Generally, the only new firms that emerge are those created from the staff of an existing company. If the importance of image and the need for nothing to go wrong are added to the equation, the barriers for new firms seeking to enter the market are raised even higher. The last thing a high-profile band would want is fans to be injured at a concert or 'roughed up' by security. Such events feed very quickly through the media, which can result in the kind of publicity most bands do not want. Consequently, groups and promoters look for the best security possible and are generally reluctant to trust a security firm that has little experience of pop security. Hence entrance to this sector is protected by a 'glass wall', which could almost be described as informal regulation. Such barriers to the provision of stewards at other events are not so strong.

Barriers to entry to the door supervision sector for contract firms depend on the geographical area. Regulations are more stringent in areas covered by door supervisor registration schemes, but in those where there is no such scheme barriers are virtually non-existent. A company can be established without a guard dog, a uniform or even headed notepaper, although some might argue that a strong physical presence is essential. This virtual absence of requirements, combined with the policy of many clubs and pubs of hiring the cheapest, results in similar consequences to those in the static guarding sector — a downward spiral in standards.

The common bond that unites most individuals working in this sector is the part-time nature of their work. Door supervisors are employed at weekends and in the evenings, and many are part-time simply because they are only required for a few hours per week. In some instances, door supervisors employed by a contract firm may work for a number of clubs, and in other security activities as guards, but generally this is not the norm. Most door supervisors have a day job or role and then work at clubs and pubs in their free time.

Similarly, most stewards are part-time, irrespective of whether they work at concerts or sports events. Again, by the very nature of the venues where they work, labour is required in large numbers, but at limited times — a football club, for example, may require a couple of hundred stewards every other Saturday, and the promoter of a pop concert might need a few hundred stewards for no more than a dozen or so concerts over a period of a few weeks every year. There are more senior stewards and managers who may work full-time, but they are in a minority. Again, there is no research to draw upon to illustrate this, other than the obvious pattern of demand for stewards at such events. There are also stewards who are voluntary, and who may offer their services on only a few occasions a year. These might include stewards on a march or demonstration by a group such as CND or a trade union. These could be classified as security stewards but, because of their occasional and voluntary nature, they have been excluded from this analysis.

Women are increasingly used by nightclubs and pubs as door supervisors. A survey by MORI found that two out of five clubs favour women (*The Observer*, 26 November 1995). They are increasingly used for their 'calming effect' and because they can legally search both women and men (whereas men can only legally search men). Nevertheless, women can be as subject to physical violence as men. One female door supervisor interviewed claimed, for example, that many men had 'no sense of valour' and that she had had her nose broken several times.

Regulatory structure

The regulatory arrangements for stewarding and door supervision services are very different from those in any other part of the industry. Voluntary self-regulation is virtually non-existent, as most firms engaged in this sector are not members of either of the two main manned guarding trade associations. Only in November 1999 had the first British Standard for this sector been published (BS 7958). In 1996, however, two associations were formed to represent individual door supervisors, the National Association of Licensed Door Supervisors (NALDS) and the National Association of Registered Door Supervisors (NARDS) — although neither would claim to be regulatory bodies. They have had very little impact on this sub-sector.

The door supervision sector displays some of the worst problems associated with any sector in the private security industry (HAC, 1995). Given this vacuum in standards it does not seem surprising that many local regulatory schemes have emerged. Almost all are based upon conditions attached to the Public Entertainment Licence of a nightclub or pub, requiring that if door supervisors are employed they must be registered with a specific body. Registration with that body usually involves the individual meeting a minimum character requirement, attending a training course and abiding by a code of practice — although the schemes vary around these criteria (Home Office u.d.). Most large towns and cities now have such a scheme. The schemes in operation can be distinguished by a number of criteria, including whether they are compulsory or voluntary, and whether there are basic character requirements and minimum training standards; they can also be distinguished by which organisation controls the scheme.

Many of the schemes have been very positive in reducing crime. In early 1992 a door registration scheme was introduced covering the Newcastle-upon-Tyne city centre. During 1991 there had been 91 recorded incidents by door supervisors of 'gratuitous violence' or drug-related offences. In the first 18 months of the scheme there were only 38 such incidents, most of which were (mainly minor) assaults (BEDA, 1995). A number of problems, however, have also been identified with such schemes. First, there are many areas without any registration scheme. Second, the schemes do not cover managerial staff if they are not door supervisors, and hence an individual with an undesirable character could control 'legitimate' door supervisors. Third, there are problems of reciprocity: a door supervisor registered in one area will often have to re-register in another. Finally, some of the schemes cover occupations that should perhaps be excluded; for example, security stewards at pop concerts at the large indoor arena in Newcastle-upon-Tyne must be registered as door supervisors (whereas at a similar arena in Manchester they are exempt).

Stewards are subject to even less regulation than door supervisors, as there are no schemes setting character or training requirements for them. However, a pop concert requires a Public Entertainment Licence, and sports stadiums also require licences. These licences often have conditions attached with regard to stewards. A sample licence for a concert at Wembley Stadium had conditions requiring stewards to be trained, briefed for duties, distinctively dressed, controlled from a central point and properly supervised. Such conditions usually relate to the two codes of practice relevant to these activities: the 'Purple Guide' (Health and Safety Commission, Home Office and Scottish Office, 1993) and the 'Green Guide' (Department of National Heritage, 1995). Nevertheless, there are no mandatory requirements for stewards to be trained to a minimum standard nor for character assessment. Some stewards — as mentioned above — are covered by a door registration scheme and must be registered under it.

In one specialist area of door supervision there is a national regulatory framework. Security staff employed at casinos are in many ways similar to door supervisors. They undertake access control, maintain order within the casino and have responsibility for emergency procedures. They also often have the more sophisticated role of preventing and detecting gambling malpractice. Such security staff are required to gain 'certificates of approval' from the Gaming Board, as stipulated by the *Gaming Act* 1968. Those who have criminal records or are otherwise considered undesirable persons can be refused a certificate, and thus denied the opportunity to work in this sector (Home Office, 1994a). Private legislation was also passed in 1995 with the *London Local Authorities Act*, which sets out the regulatory requirements for the licensing of door supervisors by participating councils. The provisions allow for registration of door supervisors, minimum training, grounds for refusal of registration, and enforcement, to name but a few.

Close protection services

One of the most coveted occupations in the private security industry is the role of bodyguard, or 'close protection' as it is often known. Films such as 'The Bodyguard' have glamorised the role to such an extent that discussions with schoolboys often reveal that their ultimate aspiration is to become a bodyguard. The activities of this sub-sector are a specialised part of the static manned guarding sector, essentially involving the protection of individuals rather than property or valuables. It may be related to the static guarding sector, but is composed of a distinct and specialised set of firms and individuals undertaking unique tasks, which most security firms are not capable of doing. Nevertheless, many static manned guarding companies are increasingly offering services in this area at the bottom end of the market. Close protection services can be defined as:

> The provision of private protective services — whether by an in-house organisation, contract company or combination of both — for individuals vulnerable to attack, by specially trained personnel.

Clearly, most close protection officers are also engaged in all the four functions — protection, crime prevention, loss prevention and order maintenance — that illustrate 'private securityness'. Close protection services are required by a variety of people, ranging from royalty, politicians, businessmen, celebrities, judges, witnesses, and scientists involved in animal experiments, to vulnerable groups such as women who have to work at night in violent areas. The last of these represents a growth area in recent years, due to the increasingly violent nature of society. Many female doctors who are on late night call employ bodyguards to protect them (*The Sunday Telegraph*, 18 February 1996). Some clients may also use a bodyguard as an 'accessory' to enhance their image, as many pop stars do. Ultimately, anyone who is vulnerable to attack and/or wants to portray a certain image can hire close protection services if they have the necessary financial resources.

The type of potential threats these different groups face varies considerably. One of the most serious threats for political figures, heads of state and other celebrities is being physically attacked. A more common threat for businessmen is kidnapping, and business executives working for rich multinational companies are particularly at risk. In 1994, according to Control Risks, some of the worst countries were Columbia, with 1,768 annual kidnappings, Pakistan with 864, Brazil with 416 and the Philippines with 231. In Europe the riskiest areas were Italy with 53, followed by Russia with 22 (*The European*, 2–8 June 1995). The former Warsaw Pact countries are also becoming particularly dangerous areas for potential victims of kidnapping, with the growth there of organised crime (Handelman, 1994).

Nearly all public figures are also subject to less serious threats such as assaults, harassment by obsessed fans or simply being 'roughed up' by over-enthusiastic ones. Demi Moore was bitten by a crazed fan (*The Daily Mirror*, 29 November 1995), and Madonna recently had to face a fanatical admirer who had attempted to break into her house. There are many others in the public eye who have been stalked. These risks frequently require the 'order maintenance' skills of close protection officers.

Threats to public figures originate from a variety of sources. Terrorists have been some of the most prolific assassins. In the UK, terrorist groups have targeted prominent individuals. Organised criminals, particularly the new gangs emerging in the former Soviet Union, have indulged in assassination as well as kidnapping, serious assaults and harassment (Handelman, 1994). Numerous mentally disturbed individuals have also posed a threat in the past: John Lennon's assassin, Mark Chapman, is a good example. Potential risks may also arise from over-zealous fans or supporters; the large numbers of frenzied fans some pop stars attract can lead to serious problems if they remain uncontrolled.

This situation has spawned what is known as the 'close protection industry'. Those involved range from specialist military or police units, in the public sector, to specialist security companies, general security firms, and individuals employed simply for their brawn. When there is a serious threat in the UK, public protection is generally given by a specialist police or military unit, or left to the ordinary police. Therefore, the demand for private close protection in the UK is relatively limited. However, close protection is not generally offered to those who travel abroad to high-risk areas, or to those who face less serious threats within the UK. Given these factors, the main market for the UK private sector is for celebrities in this country and abroad, businessmen travelling to high-risk areas overseas, and foreign governments, organisations and individuals who seek to hire British expertise. There is also a growing market for ordinary people who work in areas of high violence, such as the doctors mentioned earlier. In this context, it is difficult to estimate the size of the sector, although, a figure around 1,000 bodyguards is frequently mentioned.

The private sector firms specialising in close protection services in the UK include Control Risks, Saladin, Task International and Winguard. These firms specialise in the more serious end of the market, namely businessmen travelling abroad and foreign governments. Organisations such as Showsec (International) specialise in the close protection of celebrities such as pop stars and actors. There are also many individual bodyguards who offer their services directly through specialist publications, and some are employed directly.

The main services that are provided by these firms are bodyguards to escort and drive the client. When escorting the client, the role includes walking in close proximity, watching for potential threats and intervening to avoid a threat, or addressing it if one occurs. Depending upon the risk and what the client is prepared to pay, the number of bodyguards will vary. Some are also protective drivers, whose skills include defensive driving, which seeks to avoid dangerous situations; evasive driving, which seeks to escape the threat; and offensive driving, where the vehicle itself is used as a weapon (Shortt, 1995). Depending on the risk, the bodyguard may also be required to search for explosive devices and bugs, use communications systems, use weapons and be competent in close quarters combat.

Some firms involved in close protection also offer more specialised services. For instance,

Silverwing Consultancy Services offers personal security audits and defensive surveillance. The former involves a detailed audit of a client for his/her exposure to risks and for advice on how they can be reduced. Defensive surveillance involves training clients to realise when they are under observation, as most terrorists and criminals keep their potential targets under observation to identify the best time and place to attack.

There is a common perception of a bodyguard as a 250lb bullet catcher, but to perform the role effectively a bodyguard requires skills such as intelligence, alertness, courteousness, tactfulness, diplomacy, fitness, the ability to mix with a wide variety of different people, smartness in appearance and the ability to blend into the background. Many of these skills can only be developed through extensive training (Slaughter, 1995). Indeed, some bodyguards are also trained as butlers, and drivers so they can take on double roles for their clients (*Birmingham Post*, 30 December 1992). It is these demanding requirements and the importance of the tasks involved that make entrance to this profession at the top end of the market very difficult for those who have not undergone extensive training and/or who lack years of experience. This means that the most common background for bodyguards in the UK is service in a public body such as a specialist police or military unit involved in close protection. There are, however, individuals from ordinary security backgrounds, such as bodybuilders and door supervisors, who have undergone a specialist training course and manage to secure work in this area — but not many.

There are many bodyguard training courses available for those who seek a career in the profession; these range in duration from a weekend to six weeks, and dramatically in quality. Some of the courses that offer comprehensive training include the International Bodyguards Association (IBA), Task International, and SITO is also developing qualifications in close protection. At the bottom end of the market there are also martial arts courses, which are often promoted as bodyguard courses. There is a growing market in bodyguard courses but many are of a poor standard and highly expensive. They also tend to build up unrealistic hopes among the participants that they will secure work as bodyguards. Nevertheless, increasing numbers of people want to enter the profession, given the high rates of pay. The large companies involved in the quality end of the market will generally pay a bodyguard between $225 and $500 a day (Owen, 1995). At the bottom end of the market a bodyguard for an individual such as a female doctor on late night call may receive no more than the average security guard. A very rough estimate of the number of bodyguards, based on interviews with those from this sector, would be around 1,000.

Regulatory structure

There is no statutory regulation of the close protection sector in the UK. Given the international nature of the demand for bodyguards' services at the top end of the market, such regulation might prove difficult to establish. However, many reputable bodyguards join IBA, which is a respected professional association, and some are also members of the International Institute of Security of IPSA. A number of reputable companies in this sector are members of the main trade associations of the private security industry — for instance, Task International is a member of BSIA and IPSA, while Winguard is a member of IPSA. Nevertheless, there are many individuals, firms and training organisations that are outside the scope of any regulation. This is a cause for concern in view of the importance of the close protection role and the examples of poor standards of operation within this sector.

Conclusion

This chapter has considered what is commonly known as the manned security services sector. It has explored this sector under four main parts: the static manned guarding sector, CIT, door supervision and stewarding, and close protection services. It has illustrated the very different characteristics of these four parts. Overall, the emerging differences between these sectors and sub-sectors again show the process of fragmentation in the private security industry. The following chapters, on other sectors of the industry, will further illustrate this process.

Chapter 7

Private Sector Detention Services

Private-sector involvement in providing detention services is not a new development. As chapter 3 has illustrated, private initiative has been involved in the delivery of punishment in the past. However, private-sector involvement re-emerged on a larger scale during the 1980s and 1990s, with the provision of detention services in a variety of guises in Australia, Canada, New Zealand, the USA and the UK (Harding, 1997). In the USA the private sector has been involved in a whole range of activities at the 'hard end' of the detention sector. These include the design, construction, management and financing of prisons, the escorting of prisoners, the use of prison labour, and the management of prisons and detention facilities. They also encompass the supplying of specific services to prisons, such as catering, education and health care. The private sector has also even offered to operate the entire prison system in the US state of Tennessee (Shichor, 1995).

The private sector has also been involved to a major extent in the 'soft end' of the penal system, though this has tended to encompass non-profit-making activities such as supervising offenders on probation, undertaking community work, overseeing the electronic monitoring of offenders under curfew and operating bail hostels. Indeed, in the USA the 'soft end' of the market (except for children's homes) is largely the exclusive preserve of the non-profit-making sector (Ryan and Ward, 1989). With the exception of experiments in electronic monitoring, the limited private sector involvement in the UK has tended to be restricted to non-profit-making organisations. For this reason only the electronic monitoring aspect of the 'soft end' of the penal system will be considered in this chapter.

The extensive private sector involvement in the provision of a wide range of detention services in Australia, Canada, New Zealand, the USA and the UK has led Harding (1997) to describe the phenomenon as 'Anglophone'. Such is the extent of private-sector involvement, and its power to shape public policy, that Lilly and Knepper (1992) argue that an 'international corrections-commercial complex' has emerged. Whatever the merits in these arguments, in the UK alone detention services are big business. We estimate the annual market for privately managed prisons, prison escorts, immigration detention, escort services, and electronic monitoring to have been worth over £150 million in 1998, employing around 6,000 people. (These figures have been arrived at by adding together estimates for the different parts of this sector, shortly to be discussed, and they vary in their accuracy; they exclude the private sector's supply of goods and services, such as locks, vehicles and security equipment, to the public prison sector).

This chapter will consider the private sector's involvement in detention services. It will first consider immigration detention, and second the sector's larger and more salient involvement in prison and escort services. Its growing role in electronic monitoring services will then be evaluated.

Private sector immigration detention services

The growth in worldwide travel has provided new opportunities for migrants who seek a new life in a country with greater economic opportunities. Increasing numbers of people are attempting to emigrate to the industrialised world, using both legal and illegal means. As a consequence of these developments increasing numbers have to be processed by immigration services. Some of these migrants are detained during this processing, while their case is being considered or before their expulsion from the country. In the UK the Immigration and Nationality Department (IND) is responsible for these functions and, in certain cases, also has to escort some of those immigrants and asylum seekers, who are being expelled from the UK, back to their country of destination. This is to prevent them from attempting to return to the UK. The private sector in the UK has assumed a prominent role in these areas of detention and escorting.

Immigration detention centres

Detained immigrants and asylum seekers are kept in public institutions such as prisons and police cells, as well as in immigration detention centres (IDCs) run by private contractors. The IND uses a number of prisons for detention purposes, of which Haslar, run by the Prison Service, is the largest (*Hansard*, 5 July 1994, col 122). The numbers held in prisons under the *Immigration Act* 1971 has gradually increased over the last 10 years. On 30 June 1983, 101 immigrants were detained in prisons but on 30 June 1993 the figure had reached 429! During 1993 alone there were a total of 1,837 receptions in prisons of persons held under the *Immigration Act* 1971 (Home Office, 1995b). Police cells are also used for detention purposes, but this is usually a short-term measure before the individual is removed from the UK or placed in a prison or IDC (National Audit Office (NAO), 1995).

The decision to use private security companies for IDCs was taken by the Conservative government in August 1970, when Securicor was awarded the contract for the Harmondsworth IDC. Securicor held the contract until December 1988, when it was replaced by Group 4 Total Security, which has since been replaced in turn by Burns (International) Security Services. This contract also includes Manchester Airport and provides for the inland guarding and escorting of detainees. Group 4 holds another contract, for Campsfield House, and Wackenhut UK Ltd holds one for Tinsley House.

Detainees are distinct from prisoners as they are held without charge, trial, conviction or sentence. When they enter the detention centre they do not know how long they will be there; it could be for days, weeks, months or even a year or more — although most are only detained for a few days. In 1993, 75 per cent of those detained under deportation powers were held for less than two weeks, and only 10 per cent were held for more than three months. For illegal immigrants, the figures were 83 per cent and 7 per cent respectively (IND, 1994).

These differences tend to give credence to the argument that detention centres are not prisons. Group 4 describes Campsfield House as a 'secure hostel', and one of the original reasons for hiring a contract security company was that the use of police or prison officers would be seen as too oppressive for non-prisoners (South, 1988). However, no one was capable of explaining to HM Inspectorate of Prisons (HMIP) what the difference was between a 'secure hostel' and a prison (HMIP, 1995). Ultimately the detainees are held in an institution against their will, just like prisoners. There are important differences between the two, but in reality they are two sides of the same coin in the sense of being different forms of detention.

The private IDC sector is a tiny part of the private security industry and is also very small in comparison with the private prison sector; estimating its size, however, is difficult given the cloak of 'commercial confidentiality' surrounding its costs. The number of security staff employed at detention centres can be estimated at around 200. The costs of IDCs and of in-country escorting were estimated at £11.35 million during 1994–95, and this excludes the costs of HM Prisons and police cells (*Hansard*, 26 June 1995, col 493–4).

Immigration escort services

Private security firms are also often used to escort those who have been expelled, or refused entry, on their flights home. They tend to be used for an in-flight escort when a police escort is not available or would be too costly, or when the destination might place a public official in danger (Amnesty International, 1994). The role often involves the use of restraints and physical force on detainees. It is this role that has led to many accusations of the use of excessive force, and excited much publicity, though the size of this sector is very limited — there were only 167 individuals escorted to their country of origin by private security firms between April and December 1993 (*Hansard*, 1 February 1994, col 646).

Regulatory structure

In comparison to the volume of legislation regulating private prisons and prison escorts, there is virtually none for IDCs and the immigration escort sector. Indeed, when Securicor was initially contracted for this work the only legislation that could be cited for justification was section 13 of the *Commonwealth Immigrants Act* 1962. This states that 'any person required or authorised to be detained under this Act may be detained in such places as the Secretary of State may direct' (cited in South, 1988: 107). However, since this period further legislation has been passed, which does impact on IDCs. The *Immigration Act* 1971, as modified by the *Channel Tunnel (International Arrangements) Order* 1993, gives powers to detain. The *Immigration (Places of Detention) Direction* 1994 lists those places where immigrants can be detained. At the time of writing, a similar structure of regulation for private prisons is in the process of being created by the *Immigration and Asylum Act* 1999, as set out in the White Paper *Fairer, Faster and Firmer — A Modern Approach to Immigration and Asylum* (Home Office, 1998).

Until the above legislation was passed, however, there was no statutory framework setting minimum standards of recruitment, selection and training for staff. The powers of officers employed by detention centres and escort services were ambiguous, and there was also no statutory framework that set minimum standards for the operation of security firms. No statutory complaints procedure comparable to the one operating in prisons existed, and while private prisons have regular inspections by HMIP and visits by Boards of Visitors, the procedures for IDCs were less rigorous and more ad hoc. A Board of Visitors was only introduced for IDCs in 1989, after a report by HMIP (Home Office Press Notice, 12 September 1989). They had operated for 19 years without such a visit! These and other weaknesses in the accountability structure of IDCs (Joint Council for the Welfare of Immigrants (JCWI), 1995) made the introduction of the new statutory framework under the *Asylum and Immigration Act* a welcome move.

Private prison and prison escort services

During the 1990s the Conservative government hardened its policy on crime in the belief that 'prison works' and, as a consequence, the number of prisoners increased substantially. During 1990–91 the prison population was 45,200, rising to 49,500 during 1994–95 (HM

Prison Service, 1995). On 14 January 2000, the prison population stood at 63,057 (http://www.hmprisonservice.gov.uk/Statistics/stats2.asp) and is projected to continue to rise. The 1980s and 1990s also saw a government policy of privatisation, through returning nationalised industries to the private sector, encouraging the contracting-out of public services and developing a private sector ethos in public bodies. Given the public's hostility to tax increases, particularly for prisons — and the costs of prisons are substantial — it is not surprising that the Conservative government turned to the private sector to provide the extra prison places required at minimum cost, and pursued a policy of privatisation throughout the service.

In England and Wales the Home Office is responsible for prisons through HM Prison Service. The Director General is head of this service and reports to the Home Secretary, who is responsible for policy. The Prison Service has around 130 establishments, which in 1994–95 employed a staff of around 40,000, with a budget of £1,597 million. The devolved Scottish government (previously the Scottish Office) is responsible for prisons north of the border, through the Scottish Prison Service. The responsibilities of the Northern Ireland Office are discharged through the Northern Ireland Prison Service. At the time of writing, all privatisation, except for one prison opened in Scotland in 1999, had been in England and Wales; hence this will be the focus of this chapter. The Home Office has pursued three strategies towards privatisation of the prisons sector: contracting-out the management of public prisons to private companies; using the private sector to design, construct, manage and finance (DCMF) prisons; and contracting-out prison escorting.

The private prison sector

Located between North Cave and South Cave in the beautiful Humberside countryside lie HMP Easthorpe and HMP Wolds (NAO, 1994). The prime difference between them is that the former is a conventional modern prison and the latter the first of a new type of private prison. When the Wolds, operated by Group 4 Prison and Court Services (PCS), opened in April 1992 it marked the beginning of a new and controversial sector of the private security industry. Wolds was followed in May 1993 by Blakenhurst (United Kingdom Detention Services), in June 1994 by Doncaster (Premier Prison Services) and in December 1994 by Buckley Hall (Group 4 PCS), although at the time of writing, Group 4 had lost the contract for Buckley Hall to an in-house Prison Service bid). Together these prisons employ about 1,100 custody officers, and the contracts amount to a total of some £35 million per year.

The involvement of the private sector was further expanded in October 1993, when the Home Secretary announced proposals for DCMF prisons. The following month, the Prison Service placed advertisements in the Official Journal of the European Community inviting expressions of interest from consortia interested in DCMF prisons to be submitted by February 1994. Ten consortia expressed an interest, and 6 consortia and companies were invited to tender for the first 2 prisons, one at Fazakerley, Merseyside (HMP Altcourt) and the other at Bridgend, South Wales (HMP Parc). The former contract was won by a consortium of Group 4 PCS and Tarmac and the latter by Securicor, Siefert, and W S Atkins. The contract to run Lowdham Grange was then awarded to a consortium headed by Premier Prison Services. These three prisons employ around 1,000 staff, and the contracts over the 25 years set are worth in total nearly £643 million, or crudely around £25 million per year. Further contracts have been put out to tender, and more DCMF prisons are planned.

DCMF prisons are very different in nature from the other privately managed prisons. As their name suggests, the consortium is responsible for designing, constructing, managing and financing the prisons. It must therefore assemble not only a specialist security company to manage the prison, but also architects capable of designing prisons, construction firms who can build them and a financial institution willing to lend the required capital. The prisons are owned by a private consortium for 25 years. The Prison Service only pays when the consortium starts operating the prison. Such payments are made by the Service buying places in the prison. DCMF prisons therefore represent an extension of the private sector's influence.

Private prisons are staffed by Prison Custody Officers (PCO), which is a statutory role created in the 1991 *Criminal Justice Act* (CJA) as amended by the 1994 *Criminal Justice and Public Order Act* (CJPOA). To become a PCO an individual must be certified by the Prison Service, must pass its vetting and undergo approved training. PCOs at Group 4 prisons undergo a seven-week training course lasting around 270 hours. The course covers subjects ranging from the role and powers of a PCO, to first aid and advanced control and restraint. Minimum pay and conditions are not specified by the Prison Service and are left to the contractor to set in relation to attracting the appropriate calibre of staff from the local labour market. The pay and conditions are generally lower than for a public sector prison officer, although in some cases the starting salary is slightly higher.

The private prison escort sector

In April 1993 Group 4 began operating the East Midlands and Humberside Court escort contract, and became the focus of national attention. Between 5 April and 8 June 1993, nine prisoners were lost and each escape was reported extensively in the media, particularly in the tabloid press. The losses sparked off a new repertoire of jokes about Group 4 and the company was compared to the 'Keystone Cops'. However, closer examination reveals that firstly, Group 4 should have been better prepared for the potential bad publicity attendant on any mistake in this new and controversial sector, and that secondly, poor management of prisoners was not at issue — research since this period has indicated a 40 per cent reduction in escapes since Group 4 took over from the Prison Service and police at this location.

Prior to contracting-out, prison escorts were undertaken by the police and Prison Service — a task they continue to perform for most Category A prisoners. The Prison Service supervised transfers between prisons, and from prisons to Crown and magistrates courts, using contract-hire taxis and coaches. The police used their own vehicles to transport prisoners from courts to prison (*Hansard*, 5 May 1993, col 75). This system was expensive, both in the duplication of services and in the time of highly paid police and prison officers. To reduce costs, the government decided to contract-out the transportation of all prisoners except the most dangerous — Category A — to the private sector. It was argued that a higher quality service at lower cost could be achieved through the use of the latest technology, coupled with staff dedicated to this function and paid accordingly. Despite the poor publicity, the cost of prison escorts under the East Midlands and Humberside contract operated by Group 4 in 1993–94 was £8.1 million. The cost for the Prison Service and police in the same region at 1993–94 prices was £9 million, which excluded overheads, vehicle costs, catering and medical support (*Hansard*, 5 May 1993, col 76). Since the East Midlands contract was awarded, all areas in England and Wales have been contracted-out. All contracts are listed in the table on page 74.

Table 3. The private prisoner escort services sector

Area	Company	£	Staff
East Midlands & Humberside (A7)	G4 PCS	£10m	288
Metropolitan (A3)	Securicor CS	£19m	857
East Anglia (A4)	G4 PCS	£9m	350
Merseyside, Manchester & North Wales (A6)	G4 PCS	£14m	550
North England (A8)	G4 PCS	£8m	300
South West & South Wales (A1)	Reliance CS	£9m	350
South & South East (A2)	PPS	£9m	390
West Midlands (A5)	PPS	£8m	312
Totals		**£86m**	**3,397**

Sources: Group 4 PCS press notice; *Hansard*; & Personal Communication.

Notes:

1. (A) = **Area**; **£** = the approximate annual value of the contract; **Staff** = the number of staff employed on the contract.

2. G4 PCS = Group 4 Prison and Court Services; Securicor CS = Securicor Custodial Services; PPS = Premier Prison Services; and Reliance CS = Reliance Custodial Services.

3. All figures on the value of contracts have been rounded to the nearest £ million.

4. The number of staff and prisoners at each institution may have changed since these figures were compiled during 1995-96.

Prisoner escorting is carried out almost totally in modern cellular vehicles. For the East Midlands and Humberside contract, Group 4 used 70 vehicles, of which 57 are cellular, 3 personnel carriers and 10 saloon cars (*Hansard*, 5 May 1993, col 76). The staff responsible for prison escorts must also be PCOs and go through a similar certification process to those working in private prisons. The difference is the training that they undergo. In the case of Securicor this is a course over 200 hours long, which includes the legal framework, prisoner management, security, supervision and control, and general subjects (*Hansard*, 22 June 1994, col 171). As with private prisons, pay is a matter for the contractor; generally staff receive a salary of £11,000 to £14,000, which is slightly less than for those working in private prisons.

Contracting-out procedure

The contracting-out procedure is regulated by the Public Services Contracts Regulations (SI 1993/3228) and is broadly the same for both prison escorts and private prisons (apart from DCMF prisons). First, an advertisement is placed asking for expressions of interest in tendering for a contract. Then a selection of companies are formally invited to tender. Once the tenders have been received they go to an Evaluation Panel (EP), which will normally have representatives from prison governors, ACPO, the courts, HM Inspectorate of Prisons, and specialist consultants on total quality management (TQM) and procurement. The EP then makes a recommendation to the Prison Board, which takes the final decision.

With DCMF prisons, there are a small number of differences. First, a rigorous examination of the designs takes place at the evaluation stage. Second, because of the additional need to finance the prison, consortia are first awarded 'Preferred Bidder' status. The contract is only awarded when the consortia have convinced their financiers. The applications the companies submit are very expensive, particularly with the DCMF bids — some contractors claim the latter cost in the region of £1 million to £4 million to assemble. For ordinary prisons and prison escorts the cost tends to be around £500,000.

Figure 3. The DCMF prison contracting-out procedure

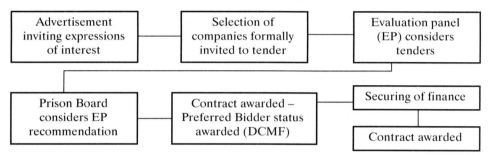

Once the consortium has been awarded the contract, a variety of standards and performance indicators are set. These are then monitored and enforced through the sector's various regulatory arrangements, which will shortly be explored. If all the contracts for the prison management and prison escort sectors are added together (excluding DCMF prisons) the annual market is worth around £110 million annually, and employs over 5,000 staff. When DCMF prisons are added, it is worth around £136 million (the annual estimate for DCMF prisons being arrived at by dividing the value of the contract by its duration, to give an annual figure).

Secure Training Centres

The 1994 CJPOA created Secure Training Centres, which were designed to be specialist prisons operated by the private sector for persistent young offenders aged between 12 and 14 years. They were proposed under the last Conservative government as a means of getting tough on young offenders. Originally five such centres were envisaged; the first, Medway Secure Training Centre in Kent, is operated by Rebound ECD Ltd (a subsidiary of Group 4) and was opened in April 1998.

Other aspects of prison privatisation

The prison service has also pursued privatisation in other areas. The education and health services are contracted out in all establishments, and catering has been contracted out in some establishments, as have prison shops. There have also been some experiments with the contracting-out of prison industries and farms, although the contracting-out of prison workshops at Codingly jail led to a doubling of losses, an internal Prison Service investigation and the ending of the Service's contract with Wackenhut (*The Guardian*, 29 January 1999).

The regulatory and accountability framework

The firms operating in this sector work within a comprehensive regulatory framework, established under the 1991 CJA and the 1994 CJPOA. In addition, privately managed prisons are also subject to much of the regulatory framework under which public prisons operate, such as the *Prison Act* 1952 and the Prison Rules.

The legislation regulating private prison escort services defines the scope of activities that can be contracted out, and gives the Secretary of State the powers to do so. The CJPOA resolved some of the problems with the CJA, and extended the contracting-out of escort services to Scotland and Northern Ireland. The CJA also created the role of the PCO and set out the powers and responsibilities of these officers, which include preventing the escape of prisoners, preventing or detecting the commission of unlawful acts, and ensuring good order. A PCO also has the power to use reasonable force, where necessary, in pursuance of his or her duties.

The legislation also sets out the framework for ensuring compliance with regulations and establishes a framework of accountability. The CJA provides for the appointment of a Prison Escort Monitor, who is a Crown servant, to keep arrangements under review and report on them to the Secretary of State. The Monitor also has the role of investigating complaints against PCOs arising from the pursuance of their duties, and breaches of discipline by prisoners under their control. The Monitor is appointed by the Secretary of State and reports to the head of the Custodial Contracts Unit in London. In addition to the Monitor, a panel of lay observers is also appointed to inspect the conditions under which prisoners are transported.

The legislation affecting private prisons follows a similar pattern to that for private prison escort services. First, the 1991 CJA, as amended by the 1994 CJPOA, provides powers for the Secretary of State to contract out the management of prisons, and the scope for doing so. Second, it sets out the powers and duties of PCOs in private prisons, which are very similar to those of PCOs working in the private prison escort sector.

However, the monitoring arrangements are slightly different in the private prison sector. For a private prison the legislation creates two statutory positions, a Director and a Controller, instead of a single Governor. The Director is appointed by the contractor, but is specially approved by the Secretary of State. The Director has most of the powers of a Governor as set out in the *Prison Act* 1952. However, investigating disciplinary charges against prisoners, conducting hearings, remitting or mitigating penalties under such charges, and ordering the removal, temporary confinement or special restraint of a prisoner are the responsibility of the Controller, who is a Crown servant. In addition, the Controller is charged with keeping the prison under review, making reports to the Secretary of State and investigating allegations made against PCOs by prisoners. The Controller is appointed by the Secretary of State and reports to the head of the Custodial Contracts Unit.

There are extensive powers that the Secretary of State can call upon if the operators of a private prison lose control. Under the 1991 CJA and 1994 CJPOA the Secretary of State can intervene if the Director loses, or appears likely to lose, control of a prison. Fines can also be imposed on the operators for less serious incidents, and they have been used in the past. For instance, in February 1994 UKDS was fined £41,166.90 as a result of a disturbance that occurred at HMP Blakenhurst (Ruddock, 1994).

All the staff of contractors working in prison escort services and private prisons must be PCOs, except for non-operational staff such as administrators and secretaries. All PCOs must have a Certificate of Approval from the Secretary of State authorising them to carry out escort or custodial functions. The certification procedure is set out in schedules 2, 6 and 7 of the 1994 CJPOA. It states that the applicant must be 'a fit and proper person', and also receive an appropriate level of training approved by the Prison Service. The Prison Service conducts an investigation into the character of all applicants, which includes a criminal records check, and all applicants must pass approved training courses. Procedures for the suspension and revocation of the Certificate of Approval are also established. The Monitor and Controller have important functions in ensuring all staff working for contractors in prison escort services and private prisons are certified. Therefore, PCOs are accountable for their actions, and malpractice or incompetence can lead to their certificates being revoked or suspended.

Private prisons are also controlled by much of the regulatory framework that applies to public prisons. They are subject to the inspections of HM Chief Inspector of Prisons. Every contracted-out prison also has a Board of Visitors established under statute. Prisoners in private prisons may also make complaints to the Prison Ombudsman, like any prisoner in the public sector. In addition to these measures, there are a number of other methods by which the operators in this sector are held to account. They are subject to Parliamentary scrutiny through questions, debates, and the inquiries of Select Committees and the Public Accounts Committee; they are also subject to the scrutiny of the Audit Commission and the National Audit Office. The sector is held particularly accountable by the media, with virtually every cause for concern being widely reported.

In addition to the standards established by the legislation and delegated legislation which contractors must meet, there are detailed standards set out in the contract between the Secretary of State and the company. The requirements in the contract range from the minimum training PCOs must undertake, the meals that should be provided and the time prisoners should be let out of their cell, to the temperature of corridors. Ensuring that the standards of the contract are being adhered to is one of the main roles of the prison Monitor or Controller, and failure to meet these standards can result in the imposition of fines. The contract is an important document in maintaining standards, although its fine detail can also be restrictive and prevent the contractor from being innovative. Nevertheless, it does represent a means to improve the standards of prisons, as Pease (1994b: 17) recently argued:

> Would it be possible to write a contract for a prison operator in such a way that the result was prisons which got better over time? ... Why should we not frame a contract which would permit a bonus for successful reform?

Unlike public prisons, private prisons have no Crown immunity and can therefore be sued (James et al, 1997). In one case this has happened, though for a ridiculous reason — a prisoner in Doncaster successfully sued Premier Prison Services for a haircut he had received, and was awarded £100 plus costs! In the future, however, the negligence of a private prison operator may result in a more serious case.

Private sector electronic monitoring of curfew orders

Experience of private sector involvement in US prisons also influenced the last Conservative government to experiment with the electronic monitoring of offenders under curfew. The electronic monitoring of prisoners in the USA can be traced back to 1983 in New Mexico. The extensive deployment of these techniques led the Home Affairs Committee to recommend, in its 1987 report *The State and Use of Prisons*, that the Home Office evaluate the US experience and assess if any use could be made of such systems in the UK. A year later the Home Office published a Green Paper, *Punishment, Custody and the Community*, which also raised the possibility of using electronic monitoring. By mid-1988 the Home Office had decided to conduct an experiment, and this began in 1989 (Mair and Nee, 1990).

Three experiments were undertaken in North Tyneside, Nottingham, and Tower Bridge, London, respectively. Contracts were awarded to Marconi in the first two areas and to Chubb in the third. As there was no legislation on electronic monitoring at the time, only those prisoners who were remanded in custody and not yet convicted could be used as subjects. The Home Office research into the experiments argued that they were too short and limited to draw any detailed conclusions. Nevertheless, the evidence of many offenders breaching their curfew, and of some also committing crimes while under curfew, gave the impression that the experiments had been a failure. This did not curb enthusiasm for the project, however, when the 1991 CJA provided the courts with the option of imposing a curfew on offenders, to be monitored using the latest electronic tagging technology. The legislation also provided the Secretary of State with the power to contract out the monitoring arrangements. Private security companies could therefore be used to keep offenders under curfew by electronic technology — in effect, monitoring their confinement at home.

In July 1995 the Home Secretary announced another experiment, to take place in Manchester, Norfolk and Reading over nine months; it was subsequently expanded to Greater Manchester and Berkshire, in November 1995. The initial contracts were awarded to Securicor Custodial Services for Manchester and Reading, and to Geografix Ltd for Norfolk (*Hansard*, 6 March 1995, col 30). The contracts were worth a total of £1.6 million (*Hansard*, 26 October 1995, col 813). Again the experiment experienced a number of problems such as tags failing to activate, breaches of curfew by offenders and a lack of telephones in offenders' homes to operate the systems. Recent research, however, has proved more positive, finding that the equipment generally worked and that three-quarters of offenders who were tagged successfully completed their curfew order (Mair and Mortimer, 1996). The scope for tagging in the criminal justice system was extended with the *Crime and Disorder Act* 1998 enabling the early release of prisoners if they are tagged. In December 1998 the Home Office announced three contracts for England and Wales, which could amount to over £35 million per year (Home Office News Release 503/98, 17 December 1998).

Conclusion

This chapter has illustrated the significant and growing role of the private security industry in supplying detention services. It has illustrated its small but longer involvement in the detention of immigrants, its large and growing interest in prisons and prison escorts, and finally the early stages of what could be a huge involvement in the tagging of offenders. In doing so we have also explored the structure, size and accountability of these different activities.

Chapter 8

Security Storage and Shredding Services

The security storage and shredding sub-sector consists of companies providing services for the storage and destruction of items of value and sensitivity. It is one of the more difficult segments of the private security industry to define because of the wide range of non-security organisations that provide similar services. Providers of security storage, safe deposits and security shredding services are clearly all security-related, as their functions include crime prevention, loss prevention and protective functions. However, there are many similar services that also have a strong element of security. A number of removal companies offer storage facilities for furniture and other household goods — items that are valuable to the owner and to thieves. Historical archives also often contain priceless documents and artefacts, which are held in secure conditions. Many banks offer a form of safe deposit service for their customers. There are also many companies that market their ordinary services as a form of security shredding, such as waste disposal firms and paper merchants. All these companies offer secure storage or disposal of valuables and sensitive items. However, the factors that distinguish these services from the private security industry are that the majority of their activities are not security-related and their staff are not usually security-vetted. A waste disposal firm may undertake to destroy a batch of documents 'securely', but most of its waste disposal activities would be related to the removal of waste, with little of a crime prevention, loss prevention or protective role. Reputable security storage companies, safe deposit centres and security shredders will also vet their staff according to security standards. This does not generally occur in ordinary storage companies or archives, though vetting is carried out in some banks. However, because bank safe deposits are part of the wider banking services and subject to banking regulation, they are not considered as part of this sector.

Security storage services

The security storage services sector can be divided into two distinct parts. The first is the storage of documents (and related items). Many companies contract out the storage of their documents and of sensitive technical specifications and plans. In the case of solicitors this involves holding their legal files. Public bodies such as the Inland Revenue and the NHS may contract out their records, and oil exploration companies do the same with their maps and samples. The second part of this sector is the storage of computer data, either in disk form or electronically, and other similar items such as video and audio tapes.

The document storage sector is unusual in the private security industry because the main factor fuelling its growth has not been crime, but rather the need to store ever increasing amounts of paperwork more economically. Many organisations operating in areas such as central London find it prohibitively expensive to store vast amounts of non-essential but often sensitive information in costly office space. They want the documents to be protected, however, where they believe that a loss of confidentiality could lead to bad publicity, to potential loss of clients or good will, and even to the likelihood of crimes being committed against their organisation. They therefore contract out the work to a company specialising in security storage. Some organisations do not have the resources to spend time storing and locating and storing files, so they hire the services of a company that specialises in this area. A number of hospitals are contracting out the storage of medical records to security storage companies to increase efficiency and to have them held in a secure location.

The motives for storing electronic information are different and, to some extent, related to crime. In this context an organisation's main fear is that a disaster such as fire, a terrorist attack or even burglary could lead to the loss of all the information held electronically. The implications of such a disaster could be devastating: for instance, for a mail order company that lost all the information on its clients such as who they were, what they had ordered and how much they owed. By the time the information had been manually re-entered (if that were possible) the company might have gone out of business. In order to prevent such disasters information is regularly backed up and stored elsewhere.

The demand for storage services specialising in electronic data protection comes from virtually any organisation that stores information on computers. The market for the storage of documents, on the other hand, usually comes from organisations that generate a large amount of paperwork. These include public bodies (like the Inland Revenue, Customs and Excise, and the NHS), large private companies, solicitors and exploration companies.

Most security storage companies will have a strategy to protect their facilities. Many will deploy a wide array of security measures, including intruder alarms, CCTV, and security officers. In addition, storage facilities will often be protected by fire prevention measures such as sprinkler systems and fire alarms. The staff dealing with the stored files will also be security vetted. Often the files come in sealed and coded boxes, with the security company unaware of what they contain; when the client wishes to use the contents, he or she merely asks for the appropriate box with the code. Not all storage is organised in this manner, as individual files may often be required. This presents more of a security risk. However, the filing systems are often very complex and many thousands may be held at a single facility. This illustrates the fact that a more serious threat to security is posed by an employee or ex-employee who understands the filing system, rather than by an outsider who would not only have to penetrate the security strategies, but also the filing systems.

With computer data storage facilities, additional measures such as providing dust-free rooms and temperature-controlled areas also have to be taken to protect magnetic data. Often the clients have their systems backed up every day, and sometimes even hourly. These disks or tapes have to be transported to the storage facility and this generally takes place in adapted vehicles that are dust-free and temperature-controlled. One company, Secure Backup Systems, has even developed a service that does not require transport as the data is transmitted electronically to its secure location at pre-determined times.

Many companies claim to offer security storage services, but there are only around a dozen that provide a reputable service; two of the largest are Hays Information Management and British Data Management (BDM). It is estimated that there are around 2,000 people employed in this sector. However, an estimate of the turnover of the market is more difficult to assess due to the lack of any research. It is also a sector where voluntary self-regulation is virtually non-existent. There are no standards, no trade association and not even a section dedicated to security storage within one of the larger security trade associations such as BSIA. However, given that the turnover of Hays Information Management and BDM is together approximately £50 million, a total estimate for the sector of £75 million to £100 million would seem reasonable.

Safe deposit centres

Safe deposit centres are essentially secure vaults containing safe deposits, which a client hires to protect his or her valuables. The Home Office Working Group (1990a: 3) on safe deposit centres, set up after the Knightsbridge robbery, defined them as:

> commercial enterprises which provide purpose built secure areas within which the general public may rent lockable compartments for the storage of valuables without needing to specify the nature or value of their property. The ultimate control of access to these compartments being kept in control of the depositor.

The first safe deposit centre can be traced back to 1875 in Chancery Lane, London. However, while safe deposit centres continued to grow throughout the rest of the world, the next major centres did not open in the UK until the 1970s and 1980s. Since this period there has been a steady growth in the number of such centres in the UK. In 1990 there were 16 companies offering safe deposit services in England and Wales, most of which were based in London (Home Office, 1990a). BSIA estimates the safe deposit market to be worth in excess of £4 million. It is not a labour-intensive industry. A typical centre might have one security officer employed constantly — hence a team of four to provide continuous cover — a receptionist and some administrative staff. However, we were informed of one centre that employed a single security officer who worked without cover. During periods when the centre was closed surveillance was left to security devices, and when the security officer was on holiday or sick, the secretary took over his role!

Many banks also provide deposit facilities, but there are a number of differences between them and safe deposit centres. First, access is often restricted during bank opening hours, a period when safe deposit centres are usually much more accessible. Second, banks rarely give insurance cover, in contrast to safe deposit centres. Third, banks have to notify Customs and the Inland Revenue of each individual who has a box; this is not required in safe deposit centres. Finally, banks are subject to their own regulations. For all these reasons deposit facilities are not considered as part of the banking sector. Similarly, hotels often provide a safe deposit service for guests to protect their valuables; such facilities are also excluded from this analysis.

There is no statutory regulation of the safe deposit sector despite high-profile calls for such action, including a report by a Home Office Working Group (1990a: 7):

> The Working Group are unanimous in their conclusion that Government regulation is necessary so that effective safeguards can be introduced.

Some centres subject themselves to self-regulation by becoming members of BSIA, which has a code of practice for safe deposit centres. The code includes standards on the vetting of managers and staff, standards of physical, electronic and procedural security, and standards of advertising. This is in addition to the requirements all BSIA firms must meet as to financial viability and insurance. However, in the latest BSIA membership list only two members were listed in the safe deposit section.

There are also substantial informal barriers to entering this market. The first is capital, as it costs at least £1.5 million to build a secure safe deposit centre. Second, it is extremely difficult to secure the high standard of insurance required — in order to achieve it a company must secure a positive reaction from an expert at Lloyds. The expert reviews a range of criteria to gauge the risk in insuring the centre. If this risk is judged as too high, and no insurance is forthcoming, this effectively puts the safe deposit centre out of business, for few would trust their valuables to an uninsured centre. Finally, the centre needs to build up the required trust to gain enough clients to make the operation viable. This is very important since if a client is prepared to put a £200,000 piece of jewellery into the hands of a company, it is essential they can trust that company to keep it secure. Despite the major informal barriers restricting entry to this market, however, a number of operators have sought to enter the industry at a lower standard in recent years.

Security shredders

Organisations produce vast amounts of sensitive information, stored in various media such as paper, magnetic tape, and floppy and hard disks. There are also many products such as counterfeit, confiscated and damaged goods that also need to be destroyed. In the past most firms left documents and goods for the removal services either to dump or to incinerate. While such practices still continue, the changing nature of crime has led many firms to employ the services of specialist security shredding firms to destroy information and goods. The firms that offer this service comprise a small, distinct and growing sector, which can be succinctly defined:

> Firms who for a fee will securely destroy valuables and stored information containing sensitive information.

There is a wide range of information that might be of value to competitors and that needs to be destroyed, for instance information on research and development, advertising and marketing material, copies of accounts, customer/client information and personal files. Hence, potential clients could be virtually any type of organisation. If the client wants to be certain that the information has been destroyed, the services of security shredders are required. Some firms have their own in-house shredding machines, but it is sometimes not feasible to shred all documents given the bulk of material to be destroyed. In-house shredding machines are not always capable of destroying certain materials or of reducing them to the required size. Therefore, these firms turn to a security shredder to undertake the service for them. Genuine security shredder firms can offer to reduce any such material down to dust.

Vast numbers of products are counterfeited. When discovered, counterfeit products are confiscated to ensure that they do not re-enter the market, and then destroyed. Some companies with very high-profile images also destroy their own damaged goods. One leading manufacturer of jeans refuses to sell slightly faulty products, as this may affect the company's image; even jeans with minor faults are therefore destroyed. Finally, some prototype goods are developed by manufacturers but never sold and, in order to prevent them from entering the market, these too are destroyed.

Companies that offer security shredding services tend to come from one of six distinct groups: specialist security shredders; large security firms; cleaning companies; facilities management companies; document storage companies; and paper merchants. Many of these groups sub-contract their work to the specialist security shredders, of which there are two types: static and mobile. Companies offering the static form of shredding provide secure sacks or bins for their client, who then fills the containers with the information that needs destroying. The security shredder picks these up when full and transports them to a secure unit. The containers are then emptied into machines that destroy the material, reducing it down to various sizes, depending on the service required (dust being the ultimate size). The security shredder then supplies the client with a certificate of destruction.

Mobile security shredding firms also supply containers but, instead of a vehicle turning up to take the information away, it has an on-board shredder, which destroys the material at the client's premises. Again, a certificate of destruction is issued on completion. Mobile shredders are rarer in the bigger cities as they need parking facilities to undertake their work, and are more common in areas with large industrial estates with ample space available. With both types of shredding service destroyed paper is then sold on to a paper merchant to be recycled. Other destroyed materials are usually incinerated if they cannot be recycled.

The structure of the security shredding market is difficult to assess because the sector is at a very early stage of development. The other major obstacle, which is a recurring problem throughout the private security industry, is the lack of detailed research. However, to enter the market as a legitimate player a large amount of capital is required — purchasing secure sacks and containers for clients, a vehicle and a secure site with a shredder can add up to an investment of over a hundred thousand pounds, though an unscrupulous operator could establish a business with a single vehicle. Some of the legitimate businesses argue that they are increasingly undercut by such operators, who claim to offer a similar service but in reality simply collect the waste and take it direct to the paper merchant, public incinerator or dump without shredding.

This can be done because there are no special legal security requirements to establish such a business. The sector has only recently sought to establish a voluntary self-regulatory framework, with the creation of the National Association of Information Destruction (NAID) in 1997. There is also a British Standard for Information Security Management (BS 7799), which states the importance of selecting a suitable shredding contractor. One such company has proposed six criteria for judging whether a firm is legitimate: first, does the firm claim to be a specialist security shredder, or some other type of company with a security shredding service? Second, are the premises where the shredding takes place secure? Third, can the firm guarantee destruction down to dust size? Fourth, is the firm affiliated to a professional security association? Fifth, does the company offer a certificate of destruction, backed by a bond? Finally, can the firm offer advice on secure data destruction? Unless a firm can answer all these questions in the affirmative, it should not be used.

Conclusion

This chapter has considered the small but distinct security storage and shredding services sector and illustrated how it can be divided between security storage, safe deposit and security shredding sectors. The chapter has also illustrated the limited nature of the regulatory structure that applies to the sector.

Chapter 9

The Professional Security Services

The term 'professional' invokes a strong debate in private security as to whether it can be applied to some of the occupations within the industry. Indeed, the *International Journal of Risk, Security and Crime Prevention* published a debate on whether security management is a profession (Simonsen, 1996; Manunta, 1996). There are certain roles in the industry where a strong case can be made. Our analysis of different careers within the industry has identified two that can be clearly regarded as professions: security consultants and professional investigators. This is because they meet to varying degrees some of the requirements that are judged to distinguish a profession, including a code of ethics, a body of knowledge, recognised professional associations and an organisational structure typical of most other professions (that is, a professional-client relationship). The extent to which these two professions meet such requirements does vary, but we argue that overall both warrant consideration as a profession — at least in the early stages of development. This chapter will now move on to consider each of the two occupations.

Security consultants

Security consultancy is a relatively new profession, although the practice of advising on security has been going on for as long as private security has existed. The emergence of security consultants has been encouraged by the rapid growth and sophistication of the industry. However, there are two problems in defining their role. First, there are many security salesmen who describe themselves as 'independent'. Second, there are many unemployed security managers between jobs who describe themselves as security consultants. Analysing the work of security consultants as a separate activity also presents problems because of the very small numbers involved. They are also difficult to place within the wide spectrum of different private security activities, as they advise on security issues from every sector of the industry. Security consultants must also be distinguished from police crime prevention officers, who also advise on security. As the former share certain distinct characteristics and are very influential within the industry, they warrant consideration as a separate body in their own right. A preliminary definition of a security consultant might be as follows:

> An individual with expertise in a particular security discipline who advises a client on security issues for a fee.

A further distinction must be made between genuinely independent consultants with no links to a company, providing security services or products, and those who do have such links or are even employed by a security firm. Moreover, it is not uncommon to find individuals employed by a security firm who describe themselves as 'independent security consultants', when clearly they are not. There are also many burglary insurance surveyors who carry out similar work by offering advice on security, but who are employed by insurance companies. They are independent of security firms, but not of other interests; these individuals could be described as quasi-independent security consultants. A further dimension can thus be added to the above definition of a security consultant:

An individual with expertise in a particular security discipline, but with no financial interest in or employment relationship with any security firm or organisation, who advises a client on security issues for a fee.

Security consultants undertake a wide array of functions which are clearly linked to three of our four core functions of private security, since they undertake crime prevention, loss prevention and protective roles. The Association of Security Consultants (ASC) list 45 identified disciplines in their promotional literature. These include: preventative surveys, electronic system specification, tender evaluation, expert testimony, contingency and disaster planning, and risk assessment. These activities would in many cases be undertaken by in-house security managers and advisers. As truly independent advisers, with no financial interests in or employment relationship with, any firm seeking to sell security products and services, they could be described as in-house independent security consultants. Indeed, most security consultants come from a background of in-house security management or advisory services for a company, although some have served in the military or police. Many security practitioners also act as in-house security consultants, as they have no management responsibilities; their role is to advise line management and they are often called 'security advisers'. There are also a number of security functions that require specialist expertise, where occupations are developing quite separately from the mainstream security industry: computer security specialists provide a good example.

Such is the increase in the sophistication of the private security industry's products and services, as well as in the risks to organisations, that it is now impossible for a security manager and adviser in a company to have a knowledge of all the relevant areas. This has resulted in greater demand for security consultants with expertise in specific areas. Because of the increase in crime, there are also an increasing number of line managers who have suddenly been given a security responsibility about which they have little knowledge. For instance, the purchase of security products and services for a school 10 to 20 years ago would have been fairly basic, presenting few problems for a head teacher. However, given the increase in and changing nature of crime, which schools have not escaped, more sophisticated and increasingly expensive security measures have become a necessity, but about which the principal is likely to know very little. Hence, he or she turns to a security consultant to provide the advice needed.

A typical example of a security consultant's work is the provision of advice to an organisation that plans to install CCTV. The security consultant talks to the client and finds out what is wanted, and the range of risks to be countered. The consultant then uses his/her knowledge to draw up a specification detailing the number and type of cameras required, where they should be fitted, and any additional equipment that might be needed. The client organisation then uses that specification document as a basis for suppliers to tender, and the security consultant might even have a role evaluating these tenders.

The number of independent security consultants, and of those with links to other interests, is difficult to assess because of a lack of empirical research. Many ordinary consultants are also hidden in the total number of security firms' employees. The ASC has around 40 members out of an estimated total of 100 independent consultants. In addition, there are burglary insurance surveyors, who work for insurance companies and also advise on security, whom we have described as quasi-independent; there are around 600 of these. Finally, there are probably several thousand consultants in security firms. If these security consultants were taken into consideration a figure of around 1,000 would be reasonable — and probably an underestimate. If one then assumed their average earnings were £20,000 per annum, this would give an approximate market value of £20 million.

Regulatory structure

Security consultants are subject to no statutory regulation; anyone can establish a security consultancy or call themselves a security consultant. However, there are a number of self-regulatory bodies that provide a degree of regulation. The main such body is the ASC, but both IPSA and BSIA have security consultants as members, although the latter only represents consultancy companies. Most burglary insurance surveyors are members of the Association of Burglary Insurance Surveyors (ABIS).

There might be no legal barriers to entry to this sector, but there are substantial informal barriers which might deter potential independent security consultants. Gaining work in this area can be very difficult for newcomers, as much of it comes through networks of security managers or other security consultants. Moreover, many companies and security managers are not willing to use the services of consultants they know little about in such a sensitive area. Indeed, some very reputable ASC members have struggled to find enough work to make a reasonable living. The security consultant community is thus a very small and closed one. This sector does not seem to suffer from the endemic problems of others, such as criminal infiltration and poor standards of operation, but this may change. Increasing problems are likely to emerge as the profession continues to grow and is seen as a lucrative occupation to enter.

Professional investigators

The terms 'professional investigator', 'private investigator' and 'private detective' are often used interchangeably, but the first of these is usually used to describe the wider investigative community, which includes both private investigators who contract for work, and in-house investigators. As this section is considering both categories, 'professional investigator' is the more appropriate title. 'Private detectives' and 'private investigators' are also often used interchangeably to describe contract investigators. Many in the profession, however, prefer 'investigator' to 'detective' because of the latter term's association with the police.

The general public's perception of a professional investigator is founded on private investigators and their portrayal in the media. Private investigators have both a glamorous and sinister image. They have been glamorised by the many fictional characters portrayed in books, television dramas and films, where they often have superior investigating skills, integrity and success compared to some of their real-life counterparts. At the same time, however, they are sometimes thought of as being snoopers, corrupt and incompetent, mainly as a result of their past work with matrimonial cases. The positive image of private investigators has been furthered through fictional characters such as Sherlock Holmes, and numerous television private investigators such as Chandler and Company, Mike Hammer, Magnum PI, Philip Marlowe, Jim Rockford, Eddy Shoestring and Micky Spillane, to name but a few. The image has been particularly influenced by private investigators in the American fictional context.

A definition of professional investigators presents large problems, given the wide diversity of work they undertake and the other occupations that carry out similar work and/or compete with them for the same assignments. When considering the problem of definition one must remember there is a large grey area of activities involving professional investigators, where an equally valid case can be made for including them within the bounds of the profession as for excluding them. We have included some of these grey areas under marginal sectors of the industry in chapter 11; others might contend that some of the occupations defined as falling within these marginal sectors should have been placed in this section, and some may argue rather that certain activities discussed in this section should have been included in chapter 11. Given the difficulties in drawing a line in this chapter, our definition of professional investigators might be taken as a starting point for debate:

> Individuals (whether in-house or contract) and firms (other than public enforcement bodies) who offer services related to the obtaining, selling or supplying of any information relating to the identity, conduct, movements, whereabouts, associations, transactions, reputation or character of any person, group of persons or association, or of any other type of organisation.

Activities of professional investigators

An examination of the activities of British professional investigators reveals a reality far removed from any glamorous fictional image. The vast majority of them work on mundane process serving, tracing missing persons who have unpaid debts, and routine investigations into fraud. Many would not know how to load a gun, let alone undertake a murder enquiry. As one writer on US private investigators has argued (Buckwalter, 1984: 9):

> Their most lethal instrument is the camera with which a private investigator documents evidence. With deadly accuracy, photographs and movie film can expose fraudulent claims and identify persons, documents, impressions, and objects.

These comments also have great relevance for British private investigators. A review of private investigators' promotional literature, discussions with investigators (private and in-house) and analysis of the literature reveals the wide variety of activities they undertake (Boothroyd 1988a, b, c; Gill, Hart and Stevens, 1996; Gill and Hart, 1997a, b; and Button, 1998). Many of these activities also merge into other sectors of the private security industry, illustrating the multi-disciplinary nature of some investigators. These activities include, *inter alia*, accident investigations, adoption enquiries, anti-counterfeiting, asset tracing and recovery, corporate intelligence services, industrial counter-espionage services, fraud investigations, matrimonial enquiries, pre-employment/employee screening, process serving and tracing missing persons. These show that professional investigators are involved in all the four core functions that define private security: crime prevention, loss prevention, order maintenance, and protection.

Some of these activities will now be considered in more depth (most of the examples relate to private investigators, but many of these activities are also undertaken by in-house investigators). Many organisations are embarrassed by large-scale frauds and do not want to involve the police as this will almost certainly lead to a prosecution and bad publicity. Therefore, many turn to professional investigators to find the culprits and, if possible, to recover the lost revenue. There is also a growing trend towards pursuing litigation for compensation, and many investigators are employed to investigate the validity of these claims. In one case an employee claimed he had a shoulder injury and pursued a £100,000 compensation claim. His employers, Manchester City Council, suspected he was faking the injury and employed a firm of investigators; they followed him on holiday and filmed him carrying heavy suitcases with his 'injured' shoulder and hauling himself from a swimming pool. On seeing the video evidence with his union official, the employee dropped his compensation claim, saving the council both legal costs and the potential compensation pay-out (*Manchester Evening News*, 24 September 1994).

Companies are increasingly subject to industrial espionage (Gill and Hart, 1997b; Heims, 1982), and many private investigators offer their services to counter this activity. In one case (Joselyn, 1994), which illustrates the problem, a firm of builders suddenly found it was losing all its bids for local authority work to a rival firm, when it had been used to winning a fair proportion. The firm did not suspect the local authority, but presumed that their offices were bugged. A firm of private investigators was called in to scan the premises for bugs, using specialist equipment, and one was found in the estimator's office (Rather than removing it, the firm then used the bug to feed misinformation on the cost of its bids!).

Private investigators are also involved in investigating firms that counterfeit various types of prestige products. In one example Carratu International was asked by a large spirits company to investigate the counterfeiting of a brand of vodka. Their investigations found 17 different counterfeits being produced in 7 different countries. Such was the problem that the client had to redesign the label using a hologram (Carratu, 1995). Many of the prestige investigative companies involved in counterfeit product investigations are also involved in due diligence enquiries, where the claims of firms and individuals are assessed to reveal if they are accurate or not. This is particularly important during takeovers and mergers of companies.

Private investigators are also employed by local authorities to compile evidence on noisy and antisocial tenants where eviction orders are contemplated. Sunderland Council used one firm that was installed next door to troublemakers to gather evidence, which was then used in court to secure eviction (*The Independent*, 20 March 1995). Some firms use private investigators to check that employees are working properly, or are genuinely sick. For example, the West Midlands Health Authority used one firm to assess how much private work its consultants were doing (*The Sunday Times*, 15 January 1995). Some worried parents are even hiring private investigators to find out what their children do when they go out for the evening. This includes investigations into whether they take drugs (*Walsall Express and Star*, 16 February 1996). House buyers are even turning to investigators for a form of due diligence enquiry, to find out what prospective neighbours are like (*Walsall Express and Star*, 22 March 1996). The hamburger chain McDonalds used private investigators to infiltrate a group of London environmentalists and gain evidence for a libel suit it was pursuing against them (Vidal, 1997). Some married and engaged women even hire specialist agencies to check if their husbands or fiancés are faithful. The agencies send a glamorous female to chat up the male to see if he gives in to temptation (*The Observer*, 17 April 1994). These are just a fraction of the wide range of cases in which a private investigator can be engaged, and many of them are also carried out by in-house investigators.

Occupations competing with and linked to professional investigators

There are also a number of occupations that carry out comparable functions to professional investigators, as well as compete with them. Some of these occupations will be briefly explored. Store detectives do not ideally fit into any sector of the private security industry. Some would identify them as part of the static manned guarding sector, while others would characterise them as private investigators (SITO includes store detectives under the manned guarding sector in its training directory). Store detectives carry out investigations in an undercover capacity, which sets them apart from the static manned guarding sector (Murphy,

1986). However, they do not fit many of the requirements of a professional and their investigations are of a much less complex nature than those of most professional investigators.

The increase in fraud in the post-war period has led some accountants to specialise in this area of investigation: such specialists are known as forensic accountants. They are relatively rare, as only big accountancy firms have a forensic accounting department, the role which can be summarised as the investigation of fraud and recovery of losses. These departments will generally have a core of specialist forensic accountants, supported by specialist investigators from a background in the police, Customs and Excise or Inland Revenue. Their work stems from two major sources: discrepancies discovered by ordinary accountants during an audit; and when the collapse of an institution due to fraud or a major fraud is identified. It is because of their specialised role in investigating fraud and other related business crimes that forensic accountants can be considered as a form of professional investigator.

Firms of solicitors may also become involved in investigations. For instance there is one firm of lawyers, Mishcon de Reya, which specialises in helping clients who have been defrauded to recover their assets (*The Financial Times*, 2 April 1996). The recovery of a large proportion of the money lost in the Brinks Mat robbery was undertaken and co-ordinated by a firm of lawyers, Shaw and Croft. However, the recovery of £22 million of the £26 million that was lost, involved a team from a wide array of occupations and disciplines, including professional investigators. Over 200 attended the party celebrating the end of the investigation, and these included: solicitors, barristers, accountants, police, private investigators, bankers and loss adjusters, to name but a few (personal communication).

The structure of the professional investigation sector

One of the easiest businesses to establish is that of a private investigator. Indeed, a correspondence course promotional pack for budding private investigators underlines the point:

> There is no other home-based business which can be started for as little money as a PI agency — and this is a fact. (The Medina School of Private Investigation, promotional material)

It may be easy enough to establish a business as a private investigator, but it is another thing to gain enough work to make a living. Much of the work undertaken by private investigators comes from informal and established networks, which range from solicitors and other agencies to insurance companies, finance houses, building societies and corporate clients. It is not easy for an aspiring investigator to receive commissions from such organisations, as the often sensitive nature of the work means they will not trust unknown investigators. New entrants who do succeed are generally those who have already had contact with such clients, either through another private investigator's agency, or in a previous career such as in the police. Many advertise in the Yellow Pages, but this brings very little business and one investigator at the 1995 ABI annual general meeting claimed 'only the nutters' would attempt to hire an investigator through this directory.

The practice of sub-contracting work is also widespread amongst professional investigators. This usually involves either one firm of private investigators contracting work to another to undertake a specific activity, or an in-house investigation department contracting to a private investigator. There are a number of reasons why an investigator might sub-contract work. The original investigator might not have the resources to carry out the task. Alternatively, there might be a requirement to conduct an investigation in another part of the country where the original firm has no expertise or resources. In certain cases a specialist investigation service is required, such as surveillance, that the original investigator could not perform as effectively. Some firms contract out when illegal or unethical activities need to be carried out, such as gaining access to the police computer or other confidential sources, telephone tapping, or bugging. Take for instance a private investigation agency hired by a company to find out if a prospective employee has a criminal record: the investigators might hire another firm, to discover the relevant information by whatever means necessary — including illegal means. The information is then laundered by the original firm, which finds the relevant press report of the conviction.

There are also a substantial number of in-house investigators employed by banks, building societies, insurance companies, retailers, local authorities, government departments and numerous other types of organisation. Three distinct forms of in-house investigator can be identified. The first is employed in an investigation department of a company or organisation — which may be part of a wider security, investigations, loss prevention or some other named department. The second is a lone investigator employed to carry out enquiries. Finally, some security managers have a range of responsibilities, one of which may be investigations.

The ease of entry into the profession, then subsequent difficulty in securing private investigative work, leads to a high turnover. In addition, there are many retired police officers and other investigators who carry out the occasional inquiry, and in-house investigators, some of whom may have a wide range of other responsibilities. These factors illustrate the immense difficulties in attempting to identify the number of professional investigators, though some individuals and organisations have arrived at estimates for the size of this sector, particularly private investigators. These estimates have ranged from a low of 3,000 (*The Guardian Weekend*, 8 September 1990) to one by Peter Heims of 10,000 (Boothroyd, 1988a). An estimate by one senior member of the ABI suggested that there were 15,000 private investigators (personal communication).

Probably the most detailed estimate of the number of professional investigators was compiled by the Institute of Professional Investigators (IPI) in 1992. The IPI estimated that there are 6,000 employed as sole operatives or in small partnerships, 8,000 engaged as individual investigators in the widest context and 1,000 in-house investigators, making a grand total of 15,000. However, there is no explanation of the research methods or evidence offered to support this figure, and it is therefore difficult to assess its accuracy. The same report also estimated the sector's value to be around £110 million, comprising £90 million from corporate business, £10 million from private individual business and £10 million from overseas business. Following discussions with senior members from this sector and taking into account the estimates above, we have come to the conclusion that around 15,000 professional investigators would be a reasonable figure. If one also used the rough estimate of average earnings of £15,000 per year and multiplied this by 15,000, this would give an approximate sector size of £225 million.

There are no dominant companies with a large market share in the private investigation industry. The sector is characterised by many hundreds if not thousands of agencies and individuals across the UK, most with a small local market. However, a group of firms dominates the large corporate investigations market; these firms include Argen, Carratu International, Control Risks, Kroll, Network Security Management and Pinkerton (now part of Securitas). It is not possible to estimate their market share. Gill and Hart (1997a), in their research on private investigators, were able to identify four 'ideal types' of private investigator. The first they call 'home based', which refers to a lone investigator working from home, usually with a low annual turnover and who is often an ex-police officer. The second type they call 'high street agency', which usually has a high street office, is a sole trader or partnership, and also tends to be staffed by ex-police officers. The third type they call the 'regional agency', which is a variation on the second but is more likely to be a limited company with branch offices and a much higher turnover. The final model they identify is the 'prestige company', which is usually based in London and specialises in high-level fraud and other specialist areas for blue chip companies. Some of the agencies also work in other sectors of the private security industry, the most common being the provision of security audit/risk services, close protection services, security officers and store detectives. There are also some guarding firms that offer investigation services.

Regulatory structure

Many private investigators are also Certificated Bailiffs, which means they are subject to a form of statutory licensing. To become a Certificated Bailiff individuals must apply to the County Court; they must be persons of good repute, with no County Court judgements against them, no bankruptcy proceedings and no criminal convictions. In addition a bond must be posted of £5,000. A Certificated Bailiff can then execute warrants in respect of Council Tax for local authorities, VAT for Customs and Excise, income tax for the Inland Revenue, and outstanding fines for Magistrates Courts. Private investigators who are also Certificated Bailiffs have thus met certain minimum standards; many also join the Certificated Bailiffs' Association of England and Wales. However, there are many bailiffs who are not certificated, as anyone can adopt the title. Some investigators also seek to gain legitimacy and respectability by gaining a Consumer Credit Licence, as this not only allows them to investigate the financial background of individuals but also involves positive vetting of themselves in order to gain the licence.

There are a number of self-regulating bodies for investigators; the largest and most influential is ABI. Private investigators who join ABI must submit an application and eventually pass an examination on issues related to private investigators. All members are subject to ABI's code of ethics, which covers issues such as respecting an individual's privacy; conducting investigations within the bounds of legality, morality and professional ethics; and verifying the credentials of clients to ensure they have lawful and moral reasons to instruct the investigation (ABI, 1995). Breaching this code can lead to a fine, and suspension or even expulsion from ABI. The other main body is IPI, which is open to the 'wider investigative community' such as in-house investigators, the police, the military and forensic investigators. IPI is essentially an academic body offering courses in investigation, although it also has a code of ethics and disciplinary procedures. IPI has also established a Registration Council open to those who cannot or do not seek to meet the criteria for other categories of membership.

ABI has only around 450 members and IPI a similar number. Many investigators also join both bodies. Given that there may be around 15,000 professional investigators, ABI and IPI are limited in their ability to institute disciplinary procedures. Members who have been expelled from the two bodies can still operate as professional investigators.

Conclusion

This chapter has considered the two distinct occupations in the private security industry that can be regarded as professions: security consultants and professional investigators. It has explored the activities, structure and regulatory framework of these professions. Together they account for a small but influential and high-profile sector of the private security industry.

Chapter 10

Security Products

The security products sector is a significant part of the private security industry. Security equipment ranges from sophisticated intruder alarm systems that utilise the latest technology, to locks that have changed very little in design and technology in the last 100 years. They range in cost from a few pence for a security seal or small padlock, to CCTV systems worth hundreds of thousands of pounds. There are literally hundreds of different security products, utilising a multiplicity of distinct technologies, produced by thousands of different companies. The sector can be broadly divided into electronic and physical security products. More specialised sub-divisions can be made between manufacturers, distributors and installers based on the type of product and at which stage of the market they are sold.

Our analysis commences with a discussion of the problems that arise when assessments are made of the nature and size of the security products sector. We then consider the electronic and physical security product sectors. This entails analysing the different security products, illustrating their function, the firms that manufacture them and how they are distributed and installed, where applicable. Finally, the chapter will explore the structures for standard setting and voluntary self-regulation.

Problems in researching the security products sector

There are a number of problems in securing accurate information on this sector. There is a lack of rigorous and widely available research. There are a number of market reports, but these are generally prohibitively expensive and not available to the public. Many of these reports also focus upon the large product markets of this sector, such as intruder alarms and CCTV, but do not consider some of the other parts, such as detection equipment. The sources that are available generally use different methodologies, even when the same product market is being considered. Therefore, the market reports and other information that will be used throughout this chapter must be taken at best as guides.

Second, the structure of this market makes it very difficult to gain accurate information on the size of the industry, as there are very few companies that restrict their activities to one security product in one stage of the market. Most companies are involved in a number of products in the industry and various stages of the market. Some companies are active in other non-security sectors, whereas others are involved in a single product, but at different stages of the market (manufacturer, distributor and installer). The difficulty in estimating the size of the market is demonstrated by the example of firms that install intruder alarms. A number of these firms install other security products and may also be involved in manufacturing and distribution. This structure creates difficulties in estimating the size of product sectors and means one must always treat such market estimates with caution. Any estimate of the leading alarm installers, for instance, must carefully assess the figures to see if they have been adapted to take account of their diverse activities across different product markets and stages of the market.

The size of the security products sector

Given the extent of the problems in estimating the size of this sector we have tried to be as conservative as possible when making our own calculations. Therefore, the following estimates should be taken as a rough guide to the relative and absolute size of the sectors. We have also erred on the side of caution inclining towards the minimum end of estimates.

Table 4. The size of the UK security products sector

Product sector	Turnover (£m)
Electronic security systems	1,590
Vehicle security products	60
Personal alarms	97
EAS equipment	250
Detection equipment	50
Security lighting	50
Locks	324
Security fences	100
Barriers, shutters, grilles, safes and screens	270
Security glass	125
Armoured vehicles	25
Other	20
Total	**2,961**

Sources: electronic security systems, MBD (1994); vehicle security products, Key Note (1993); personal alarms, Euromonitor (1989); EAS equipment, Security Management Today (1996) adapted after interviews with experts on this sector; detection equipment, interviews with experts; locks, Keynote (1996); fences, National Fencing Training Authority (NFTA) (1995) adapted after interviews with experts on this sector; barriers, shutters, grilles, safes and screens, BSIA (1999); security glass, interviews with experts; armoured vehicles, interviews with experts; other, based upon rough estimates of remaining sectors.

The security products sector market is estimated to be worth around £3 billion. Unfortunately because of the problems in researching this sector it was not possible to use estimates all from one year or to find accurate reports from every product market. Therefore, some of the information used was over 10 years out of date! In some sectors we also had to rely on 'experts' views of the size. Nevertheless the information does provide statistics — which are probably significantly underestimated — on the market size of security products sector. There are even more problems estimating the numbers employed in this sector. However, using some of the above and other reports, combined with 'experts' views we were able to make an informed calculation of around 90,000. This consists of 33,000 installing intruder alarms (Jones and Newburn, 1994); 14,000 manufacturing intruder alarms (MBD, 1994); 2,600 installing and manufacturing CCTV (information from BSIA); 2,200 in the designing, manufacturing, distributing and installing access control systems (information from BSIA); at a conservative estimate at least another 8,000 in the other parts of the electronic security sector (various estimates and calculations); 3,000 locksmiths (personal communication); 10,000 key cutters (personal communication); 5,000 fence erectors (NFTA, 1995 adapted); 4,000 manufacturing locks; and at least another 8,000 in the other parts of physical security sectors not represented in these figures (various estimates and calculations). We accept there are problems with the calculation of both sets of figures. They illustrate, however, that even with a conservative approach, when the 'wider' product sectors are included, it accounts for a significant industrial sector.

The electronic security products sector

The rise in ever more complex technology, combined with the continued growth in and sophistication of crime, has fuelled an expansion of increasingly complicated security products. This process has been intensified by the end of the Cold War and the movement of some defence manufacturers, with their high-tech expertise, into markets such as private security.

The range of electronic security products includes: intruder, car, portable and personal alarms; alarm verification equipment; CCTV; access control systems; EAS equipment; security lighting; security shredders; tracking systems; and anti-counterfeiting, detection and specialist surveillance equipment. Some of these products are manufactured and installed by firms specialising in a single area of the market. Others are produced by multinational companies operating in a wide range of sectors of the industry. For instance, manufacturers of metal and explosive detection equipment generally only specialise in this area of the security market. Other companies are involved in the production of electronic security products across the whole sector. To complicate matters, in some product markets certain companies are involved in both the installation and distribution of these products and in others some specialise in just one stage of the market. With some products, such as personal attack alarms, there is also no need for installation. Thus, an analysis of this market is very complex because of both the fragmentation and integration of the different product markets and the varying horizontal and vertical operations of companies.

Intruder alarms

The intruder alarm market is traditionally the largest segment of the security products sector. The simple intruder alarm has been transformed during the 1980s and 1990s from electro-mechanically based equipment to alarms based upon transistorised circuitry and micro-processors. There are essentially two types of intruder alarm. Remote signalling alarms are linked to the police or a central station and communicate with them when activated. Audible only alarms cause an external device such as a bell and/or light to activate when triggered. It has been estimated that there were around 810,000 remote signalling alarms fitted in 1997 (figures supplied by NACOSS). Estimates suggest that there also over 2 million audible only alarms fitted (Jordans and Son, 1992). A further indication of the size of this sector was provided by the estimate that over 250,000 security systems are installed each year (NACOSS, 1993). The market can be further divided into the commercial (that is factories, retail outlets and warehouses) and domestic sectors. The domestic sector accounts for around two-thirds to three-quarters of the total security systems fitted. In terms of market value, however, it accounts for considerably less as domestic security systems are usually smaller, simpler and cheaper than commercial products.

Intruder alarms consist of a number of components. The most important are alarm sensors, which when activated generate an alarm condition. Of the wide range of different types of sensors available some of the most common include: active and passive infrared; magnetic contacts; photo electric; radar; seismic; ultrasonic; and vibratory sources. The alarm sensors are linked to a control panel through wiring or some other form of transmission such as microwaves. The control panel when activated by one of the sensors generates an alarm condition, which depending upon the type of alarm will transmit a signal to a police or central station, either through the telephone network or some other method such as microwaves. The activation may also generate an external device to operate such as a bell, horn or light. The external device is also usually housed in a special box. The control panel will invariably have a key pad linked to it, which is used to switch the alarm on and off and to cancel false alarms. The type of component used and level of sophistication will depend upon the type of alarm required. Some intruder alarms are also now portable, such as Sentor's portable intruder detector. It resembles a cross between a missile and a buoy.

Numerous products have been developed in recent years to improve the performance of intruder alarms — particularly to reduce the level of false alarms. One of the major developments has been the growth of central stations, which will be discussed in their own right later in this chapter. Central stations have used a number of methods and types of equipment to reduce the level of false alarms. A common strategy is to monitor the pattern of alarm activations amongst different sensors. More sophisticated forms of verification equipment have also emerged. These include audio verification, where the operator at the central station can dial into the place where the alarm was activated, to listen and assess if it was a false alarm or not. In recent years verification has been improved further, with the development of visual verification systems such as TVX (product name) and Digital Video Storage and Transmission (DVST), which enable the operator to survey the place of the alarm activation.

The manufacturing sector for intruder alarms is highly fragmented. There are a few companies with relatively large turnovers, and hundreds of small and medium sized companies with relatively small turnovers and market shares. MBD (1994) estimated there were 389 companies involved in the manufacture of intruder alarms, employing over 14,000 in 1991. Many manufacture other electronic/electrical security products and some are also involved in the physical and manned security sectors, as well as activities not linked to security. Intruder alarms are also imported and many British manufacturers export their products. The structure of the manufacturing sector is further complicated as some firms only produce components such as sensors, whilst others make whole intruder alarm systems. The commercial distribution sector is equally difficult to assess because many distributors are also involved with other products, as well as the installation and manufacture of intruder alarms. Not all intruder alarms are purchased from manufacturers or commercial distributors, however. Intruder alarms for the domestic market can also be purchased from the large DIY stores such as B & Q and Do-It-All, in addition to some of the many small local hardware stores.

Intruder alarm systems can be installed by the purchaser, but the vast majority require the expertise of professionals. This is underlined by the substantial numbers employed in the installation and maintenance of intruder alarms. The installation sector has grown dramatically during the post-war period, especially when one considers that in the 1950s intruder alarms were luxuries for the rich and organisations with substantial valuables. Now virtually every business has an intruder alarm, as do many private homes.

It has been calculated that there could be between 7,000 to 8,000 firms that install and maintain intruder alarms (SITO, 1993b). Many of these firms are also involved in the installation of other security systems such as access control, CCTV and fire alarms. These figures, however, can only be taken as a rough estimate as there has been no detailed research on the number of alarm installation firms. These figures are also complicated by the many electrical contractors and local 'handymen' who perform the occasional installation of an alarm. Despite the thousands of small companies in this sector, 10 companies fit around 50 per cent of all 'approved' installations. Nevertheless, the majority of alarm installation businesses employ between 1 to 5 people, with a turnover of around £50,000 (SITO, 1993b). The Home Office has suggested a figure of 33,000 as the employment level in this sub-sector (*Hansard*, 13 February 1996: col 877), which was probably based upon Jones and Newburn's (1994) estimate of 33,755.

Many alarms are linked to a central monitoring station. In the event of an activation the operatives at the central station will assess if it is genuine. If they think it is, the police and/ or keyholder will be notified. Some central stations might also dispatch a mobile security patrol to investigate an alarm activation. Many central stations also monitor non-security alarms such as freezer alarms in frozen food stores, which are activated if the temperature starts to rise. Increasingly sophisticated technology has allowed some companies to consolidate their monitoring into one or more stations. Such centres can deal with a large number of clients. For example, the Birmingham Communications Centre of Chubb deals with over 26,000 clients and 4,000 incoming/outgoing calls per 24 hours. Most of the large installers, such as Chubb, have their own stations, which they market to their installation customers. There are also some independent of any installers, such as Southern Monitoring Stations. NACOSS, in April 1998, had 53 registered central monitoring stations under its three categories of membership (NACOSS, 1998). There have also been a number of large companies that have developed their own in-house monitoring stations in recent years, including the National Westminster Bank, Dixons and Tesco.

CCTV systems

During the 1990s there was a huge increase in the use of CCTV in both private and public domains. This growth was fuelled by the rise in crime and the perception that such systems could have a dramatic impact on curbing crime. Whatever the debate on the effectiveness of CCTV and civil liberty issues, this sector will continue to be one of the most dynamic and fastest growing within the UK industry.

There has been a rapid improvement in the technical development of CCTV in the last few decades. The first CCTV cameras can be traced to 1951 when RCA introduced the VIDICON tube, which allowed for greater picture resolution and smaller cameras. These cameras, however, were susceptible to ghosting with moving pictures. There were also problems with vibration and electromagnetic interference. These issues were addressed with the NEWVICON camera tube in 1974, which was sensitive to infrared light. Both systems, however, still failed to provide pictures of a high enough quality. The next major step was the emergence of the colour camera, although the high cost and frequent failure did not make them popular. The replacement of the tube with a 'chip' in cameras of the late 1980s, following the introduction of the Charge Coupled Device, reduced costs and increased the quality and attractiveness of CCTV. This technology increased the working life of cameras and improved picture resolution. Other developments in recent years have included the introduction of cameras with tilt and pan, enabling cameras to be moved up and down, and left and right. Some of the modern cameras can also be pre-programmed with movement patterns. The introduction of multiplexers has also allowed the simultaneous recording of a number of separate cameras at the same time by one video recorder. Developments in video recording technology have also allowed 3 hour tapes to record up to 960 hours. Finally, technology has emerged that allows moving pictures and still pictures to be transmitted down the telephone lines.

CCTV systems generally consist of the cameras and their mountings, the transmission system, a control panel to move them (if that is possible), a multiplexer, the monitor and the video recorder. There may also be other products such as Dedicated Micro's DVST. There are wide variations in the quality, size, cost and uses of these different components. The most common use for CCTV is for security purposes. It is also used for other non-security purposes such as pipeline inspection and analysing customer shopping patterns in retail units.

Most CCTV systems are located in shops, town centres and commercial premises. These systems usually require the services of a professional installer to fit them. However, there are many simple systems that do not. Some systems are small and portable and designed to be hidden. There are tiny cameras available, which some security personnel use to keep a particular area under surveillance to detect staff theft or some other alleged criminal activity. There are also mobile CCTV systems, which are fitted to trailers and can be moved to different locations.

The BSIA estimates that there are around 2,600 employed in the manufacture and installation of CCTV systems. Analysing the manufacturing sector is complicated as there are companies that only produce components of the CCTV system. Others construct whole systems, often from parts manufactured by other companies. As with the intruder alarm manufacturers, there are also firms involved in the CCTV sector that primarily manufacture non-security equipment. For instance, Hitachi are well known for their home entertainment equipment such as stereo systems and televisions, but the company also manufactures CCTV. There are also firms that claim to be manufacturers of CCTV in the UK, such as Sensormatic, but actually import the vast majority of their equipment. Most CCTV products are imported from the Far East from manufacturers such as: Hitachi, Fujitsu General, JVC, Panasonic, Phillips and Sanyo. UK based manufacturers, such as Molynx, Videmach, Shawley, Dennard and Baxall, account for a small percentage of the market.

A feature of the CCTV market is the pivotal position of the major distributors. Generally manufacturers of CCTV have little contact with end users and instead sell their products direct to the major distributors such as Norbain and CCTV Warehouse. This is because there is a fear that greater contact with end users might encourage distributors to turn to other suppliers. This power has also encouraged the manufacturers to supply the distributors with substantial discounts.

Some small firms are dedicated solely to the installation of CCTV systems. However, most CCTV installers are also involved in the installation of electronic security systems such as intruder alarms and access control systems. Some installers will also bid for a contract to supply all the security products to an organisation and then sub-contract parts out to specialists who are solely involved in installing CCTV. The description of the installation sector for intruder alarms is generally applicable for CCTV installers.

Access control systems

Access control systems could be described as little more than an enhanced lock and key (Houghton, 1994). The functions of the systems are basically the same: to secure access for specific locations to certain individuals. There has been a steady growth in the use of electronic access control systems in the last 10 years. One cannot escape the familiar sight in the high street of a bank lobby, at night and the weekend to all except customers with plastic credit and debit cards, who use them to enter and withdraw cash. Similarly many organisations now have access control systems based upon identity cards with a bar code, magnetic strip or some other form of technology. This ensures that employees can only enter those areas they are authorised to. There are also access control systems based upon other technology such as entering a code number, or even using biometric characteristics such as the voice or fingerprints of an individual. Linked to access control systems in some instances, are the audio and video entrance systems.

One could also add access control systems for vehicles, as many systems also operate in car parks to restrict access to certain vehicles with the correct pass or, in commercial car parks, to those who have paid. Access control systems consist of a barrier, which might be a door, turnstile, blocker or boom. Linked to these is usually a form of control panel, which is entered with either a card, code, voice or money to secure access. The control panel may be linked to a computer, which will determine if access can be granted and may also record all activations. The access control system may also be integrated with other systems such as intruder and fire alarms and CCTV.

The BSIA estimate that there are around 2,200 employed in the design, manufacture, distribution and installation of access control systems (personal communication). The sector can be clearly divided between access control systems for buildings and those for car parks. The market, as with intruder alarms and CCTV, is difficult to assess because of the large number of firms involved in different products as well as different stages of the market, not to mention the problem of sub-contracting. Some of the leading manufacturers and suppliers of access control systems are the large companies such as Group 4 Technology and Thorn Security, as well as more specialised companies such as Cardkey. The increasingly sophisticated technology of access control systems has led to many being integrated with CCTV, intruder alarms and other equipment. Most of the major installers of access control systems also install intruder alarms and CCTV, although there are some specialists.

The vehicle security access control market is distinct from those systems designed for people. Products in this sector include barriers that rise, sliding gates, blockers that rise from the ground, and the systems that operate these barriers. Some of these products are manufactured partly or completely abroad. Many of the motors for raising and lowering barriers are made in Italy, for instance. In this part of the access control market most of the suppliers will also install the products.

Integrated systems

Integrated systems, as their name suggests, incorporate a range of security systems. A combination might include an intruder alarm, fire alarm, access control system and sprinkler system. Literally any combination could be possible. One trend that has been identified is combining intruder alarms with CCTV, so when the alarm is activated it triggers a camera (Steer, 1995). The overwhelming majority of integrated systems are manufactured, distributed and installed by the same companies involved in the intruder alarms, CCTV and access control systems sectors.

Other security alarms

The most common type of other alarms are for vehicles. The huge rise in car crime over the last two decades has stimulated a remarkable growth in vehicle security products. Car alarms have been one of the most popular strategies used to combat this form of crime. Many alarms are installed while the car is being manufactured. Others are installed in cars by professionals after a vehicle has been purchased and some kits can be bought from shops to be installed by the DIY enthusiast. A number of alarm systems are also linked to engine immobilisers, which prevent the engine from starting when activated. The prevalence of satellite technology has enabled companies to develop tracking systems. Therefore, if a car is stolen the tracking device, which is attached to a vehicle, is activated. This enables the police to track and locate the stolen vehicle.

The fear of crime amongst some vulnerable groups has led to the increasing use of personal attack alarms. When a person is attacked or in fear of been attacked they activate the hand held alarm, which makes a very loud noise. In addition to portable attack alarms there are also panic alarms, which can be fitted in locations where people may be vulnerable or fearful of attack. A social security office may be fitted with such an alarm to protect staff from angry claimants. An old-age pensioner living in a high crime area may also be fitted with such an alarm. Alarms for old aged pensioners may also have other purposes such as calling for help in the event of an accident or sudden illness.

Electronic article surveillance equipment

The rise in retail crime and the huge costs to retailers has led to a growth in specialist retail security equipment (see chapter 13). One of the more common strategies pursued has been the installation of EAS systems. An EAS system works on the basis that all goods have a tag. This tag is activated and sets off an alarm when a product is taken out of a shop through a detector gate at the entrance, unless it has been deactivated when purchased at the sales counter. These systems work on radio frequencies, microwave or other ultra high frequencies, electromagnetic fields and acoustic magnetic fields (Bamfield, 1994). The component parts are the tags, the deactivation and activation equipment and the detector gates. EAS systems are also frequently used in other sectors such as libraries, museums and, more recently, for the tagging of babies in maternity units.

Detection equipment

Detection equipment can be divided between technology that finds weapons, explosives and drugs, and equipment that discovers 'bugs'. The use of the former type of detection equipment expanded during the early 1970s with the growth of terrorism, particularly at airports. Security products were sought by the authorities and security firms that could detect guns, weapons and explosives. Now detection equipment is a familiar sight at airports, where it is used to screen passengers and luggage for explosives and weapons. Detection equipment can also be seen at other sensitive locations where there is a strong terrorist threat, such as government buildings, military installations and pharmaceutical companies. Detection technology is also used to screen letters, parcels and other packages in these locations. Most detection equipment is based upon metal detection or x-rays. There are, however, more complex systems for explosive detection based upon nuclear technology, advanced x-ray technology, magnetic resonance methods and vapour detectors (Baldeschwieler, 1993). The equipment can be divided between portable and permanent machines. Indeed, some of the detection equipment is very bulky and heavy.

The growth in 'bug' detection equipment in recent years has been stimulated by the increasing fear from and threat of industrial espionage. This is not surprising considering over 200,000 covert bugs are sold each year (Davies, 1996). One of the leading manufacturers in the UK is Audiotel International Ltd. There are also manufacturers of these products based outside the UK. Companies such as Audiotel sell their products direct to distributors, as well as hiring them out with their own trained staff. There are also specialist shops that focus on this type of equipment, in addition to selling the bugs this equipment detects.

Other electronic/electrical security equipment

Many other electrical/electronic security products are available. As they account for a relatively small portion of the private security industry, however, they will be briefly explored in this section. Such is the pace of development of security products that new artefacts are emerging all the time. Given the size of the industry, a small obscure security product could quite easily be overlooked. An additional problem is the fact that some products have multiple uses, only one of which is security.

The use of lighting is one of the oldest security strategies. Darkness has always facilitated criminal activity by hiding the offender from public view and increasing the fear of crime. Many households have security lights in the garden, which switch on when activated by a person walking by, as do many commercial premises. There are also many buildings that have vulnerable areas permanently lit. Many organisations also have their own in-house shredders. Electronic point of sale equipment comprises systems that verify credit and debit cards, and systems that allow the sales of a cashier to be analysed also exist. Security officers who have to patrol buildings will be aware of guard control systems. These systems are usually hand held devices that resemble guns. Security officers carry these devices and use them to record the time at which they were at a certain location during their patrol.

The physical security products sector

The term physical security products has been used to describe all security products that are not based upon the use of electricity. As with other sectors of the private security industry, there are indistinct areas. Some security fencing is electrified or has electronic sensors to detect intruders. Similarly the 'Eloctro', manufactured by Josiah Parkes and Son, encompasses both mechanical lock and electronic technology. There are also grey areas, as to whether some physical security products are actually security products at all. For example a high security fence around a factory, which is tall, has razor wire on the top and is designed to keep intruders out is clearly a security product.

Locks

Despite the advent of new technology and the emergence of new forms of locks and keys in the form of access control systems, locks and keys probably remain the most common security product as virtually every adult has a key and almost every building has some form of lock. Locks and keys, however, vary immensely from simple three lever locks for internal doors, to high security five lever locks for front doors and commercial premises. They range from mortice dead locks to cylinder locks. Locks have uses on outside and internal doors and windows. There are also small cheap imported padlocks, which can be bought from the local market, and relatively expensive tough high security padlocks such as Chubb's 'Conquest Open Shackle Padlock'. There is also a large demand for locks for vehicles and a specialist market for very high quality locks for prisons and similar institutions.

Thus the market for locks can be broken down into four areas: ordinary locks; high security locks (prisons and similar institutions); vehicle locks; and padlocks. The firms involved in manufacturing locks in the UK are largely based around Willenhall in the West Midlands. Many of the companies have long histories and can trace their antecedents back over 150 years. One of the largest manufacturers is Chubb Security Plc, which was also recently taken over by one of the other leading companies, Williams Holdings. There has also been a substantial increase in imported locks, particularly from the Far East, in the last few years. The impact has been such that many UK manufacturers are reducing their workforce because of this competition. In 1996, however, according to Keynote over 4,000 were still employed in manufacturing locks (Keynote, 1996).

The way locks are distributed has changed dramatically in recent years. Traditionally they were distributed almost entirely by ironmongers and builders merchants. The advent of large DIY Superstores, however, has led to them gaining a substantial share of around 25 per cent of the market. At the same time some of the manufacturers have also moved into direct distribution (Liardet, 1995). The distribution of locks is now mainly undertaken by iron mongers, builders merchants, DIY stores and the manufacturers themselves.

Locks and safes are usually installed by locksmiths, but builders, carpenters and those in similar occupations also often install locks, particularly during the construction of new buildings. Modern locksmiths are engaged in a wide range of functions. They include the installation and maintenance of locks and safes, providing advice on physical security, opening locks and safes (where keys have been lost), and cutting keys. Some also install access control equipment and other security products. A growing specialisation is also emerging amongst locksmiths with some who specialise in safes or vehicle security. There are nevertheless still many generalists. In addition to locksmiths there are also many individuals employed in shops, particularly shoe repair shops, who offer the service of key cutting.

The nature of the sector makes it very difficult to give precise figures for the numbers employed and the market value, although there are a number of estimates. The Master Locksmiths Association (MLA) estimates that there are approximately 1,500 locksmiths who are members of their association and a similar number who are not members. In addition to the estimate of 3,000 locksmiths, there could also be over 10,000 key cutters, although it is impossible to comment on the accuracy of these figures (personal communication).

Safes, vaults and security storage

Safes are also a common security product and are often showed in films and dramas being 'cracked' by professional criminals. They are essentially reinforced storage facilities designed to protect valuables against theft and/or fire. They are used by businesses and individuals to keep money, valuables and sensitive documents. A further growth area for safe technology has been the growth of Automatic Teller Machines as these are housed in secure storage facilities. Linked to the safe family are secure boxes and cabinets, which are used to keep items that may not be valuable, but which need to be secure, such as a gun cabinet. At the other end of the spectrum are vaults and strongrooms, which are essentially room sized safes that are built for clients such as banks, building societies and safe deposit centres. Many documents held by organisations are sensitive and need to be secured out of office hours, but are not so sensitive that they need to be placed in a safe. These type of documents are usually placed in a security cabinet.

Barriers and shutters (physical protection)

Barriers and shutters have similar roles in preventing unauthorised entry to an area. Examples of barriers include bollards in car parks as well as temporary barriers, such as rolled out barbed and razor wire, and spikes on the top of walls and fences. The increase in ramraiding, as well as the more common smash and grab, has lead to an increase in the use of shutters, which are generally rolled metal or metal mesh, placed over windows and doors.

Fences

Fences have long been used as a security product either to keep intruders out or to keep people in. There are generally four types of security fences: chain link; palisade; weld mesh; and barbed/razor wire. Some security fences are also electric and give small shocks to those who touch them. Others have sensors on them which activate alarms when touched. In the USA some jails have even attempted to install electric fences, which give killer shocks of 4,000 volts (*The Guardian*, 28 February 1993). There are also security enclosures, which are often used in factories to secure a specific area where more valuable equipment is stored. Fences, however, can also have non-security roles such as the demarcation of borders, or enhancing safety on a road by preventing cars from crossing over carriageways. Electric fences are also often used to keep livestock in a particular area.

Generally a fence is specified by an architect, engineer, landowner, official or security consultant. The fence will then be ordered from a supplier (distributor) or manufacturer and then installed. Of the wide range of individuals involved in the installation of fencing, the most prominent are fencing contractors. Manufacturers, distributors, builders, and in-house maintenance staff from the organisation purchasing the fence may also be involved. These issues combined with the difficulty of identifying those only or predominantly involved in security fencing, makes estimating the numbers employed in this sector very difficult. Nevertheless, NFTA (1995) estimate there are 4,000 to 5,000 employed as specialist fencing contractors; 7,000 employed in multi-skilled teams (involved in other work besides fence erecting, such as building, carpentry etc,); and 30,000 employed in multi-skill teams in the public sector. The problem of definition is illustrated by the example of Birmingham City Council's maintenance department, which employs 650 people and has responsibility for erecting fences. The amount of time spent erecting fences, and security fences at that, would only comprise a fraction of the department's total working week.

For ease of analysis we have included all those contractors who specialise in fencing as a basis for an estimate of the number employed in security fencing. We accept that this total might include individuals for whom fencing is only one of many activities and there are also many other individuals engaged in security fencing on a part-time basis. If all these contractors are included the security fencing sector would comprise approximately 5,000 employees.

Security glass/windows

Glass is increasingly marketed and developed as a security product. There are three types of specially strengthened glass, which are used for security purposes: laminated safety glass; laminated security glass; and bullet resistant glass. Laminated safety glass must meet BS 6206 (these British Standards are due to be replaced by European Norms) and by law can only be used in certain locations. While not designed as a security product, the increased strength of the glass means that it is used as a basic form of protection. Many shops use it to prevent smash and grab raids. Laminated security glass must comply with BS 5544 and should be resistant to repeated manual blows over a long period. Bullet resistant glass provides protection against bullets and shot and is manufactured to BS 5051 parts 1 and 2. It is used in a variety of locations including vehicles for VIPs; screens in banks and similar locations; windows and buildings under threat from attack, such as a government building; and specialist screens to enable individuals vulnerable to attack to make public speeches. Bandit resistant glass is used in shop windows, home windows and cabinets in museums holding valuables. Films have also been developed, which can be added to existing glass to make them stronger and more resistant to attack, such as those of Solar Gard. Windows for homes and commercial premises are also increasingly marketed as security products. In some cases marketing is the limit of the security element. In many security marketed windows, however, there are additional features. These include high security locks, more bolts, stronger frames and tougher glass.

Armoured vehicles

Many vehicles are specially reinforced in order to perform their security roles. The type of reinforcement can be divided between discreetly armoured cars and specially converted vehicles for transporting valuables or prisoners. Reinforcement is undertaken by a number of specialist companies. Some of these companies also carry out specialist vehicle conversions for the police, ambulance and fire brigade, as well as those with no identifiable security role such as stretch limousines and hearses.

The need for discreetly armoured cars has been illustrated by several assassination attempts, for example, the attempt upon Egyptian President Hosni Mubarak in Addis Ababa, Ethiopia. Terrorists pumped his Mercedes S-class with bullets, none of which penetrated the armour and bullet proof windows (*The Times*, 1 July 1995). The main demand for discreetly armoured vehicles in the UK is from the government (for cars to transport ministers), members of the Royal family, senior military personnel, diplomats and other vulnerable groups. There is also a demand for such enhanced cars for similar personnel from foreign governments. These cars have also been purchased by media organisations for their journalists and other crew covering war zones, and companies with senior executives in regions under threat from the Mafia.

One of the leading armoured car manufacturers in the UK is MacNellie of Walsall. The company will take a car from the manufacturers, strip it bare and rebuild it with armour and bullet proof glass so that it looks exactly the same. It will usually weigh twice as much on completion and require enhanced suspension. Other optional extras include: oil slick dispensers to fire back at pursuers; vehicle anti-tamper detection systems; modified exhausts to prevent objects that have been installed from stalling the engine; speak in/out intercoms; compressed air systems to enable occupants to breathe in the event of a gas attack; concealed weapon storage; run flat tyres; automatic fire extinguishers for the engine and passenger compartments; and vehicle tracking systems. The modifications to cars can add over £150,000 to the original price. Other companies offering these services include Autoprotect Systems Ltd of Neath in Wales, as well as a number of firms in the US. Some car manufacturers, such as BMW and Mercedes Benz, also discreetly design and build armoured cars so there is no need for them to be stripped and rebuilt. However, their design and production facilities make it difficult to add specialist armour or gadgets (Nedin, 1995).

More visible examples of specially converted vehicles that have enhanced protection added are cash-in-transit (CIT) and cellular vehicles (which are used by security firms with contracts for transporting prisoners). The enhancements for armoured cars and cellular vehicles are different to those of discreetly armoured cars as the threats to these vehicles are different.

Other security equipment products

There are a wide range of other products that could be described as physical security products. The following are some of the more popular examples but are by no means exclusive. Mirrors are commonly used in retail outlets to enable staff to keep areas under surveillance. There are also inspection mirrors, which are used to examine the underside of vehicles for bombs. Another common security product in retailing are secure units next to tills, which excess money can be placed in. The expensive nature of many security products has spawned a whole series of 'fakes', which resemble the genuine article. The most common of these products are fake intruder alarm boxes, which are fixed to the outside of a building, and fake CCTV systems. There are a number of other products linked to locks and keys that have a security role. Two of the most common are spy holes and door chains. Security seals can be traced back to ancient times, when letters were secured using a wax seal so any interference with the letter could be determined. Security seals are still very common and are often used on containers on lorries, packages and doors.

A host of new security products have also emerged to deter vehicle theft. The physical range includes immobilisers, which can be fitted to the steering wheel and gear stick. Wheel clamps are used, not only to enforce parking regulations, but also by some car dealers and owners to secure their vehicles. Other security products, which have not been discussed so far, include ultraviolet pens, which when applied to paper cash show up counterfeits. Dyes have entered the market, which spray an area when an alarm is activated and cover intruders in an invisible and unique dye, which takes weeks to rub off. Signs warning of store detectives at work and security guards and/or guard dogs patrolling are also used by firms to deter criminal activity. The theft of office equipment has also inspired some firms to produce products that secure office equipment. These products range from highly adhesive tapes to physical braces. A range of security marking products are also available on the market either to visibly or covertly mark valuables. Finally, the increase in violence faced by some door supervisors has led some to purchase protective vests. Therefore, yet another market is beginning to emerge for security products that used to be the preserve of specialist police officers.

Standards and voluntary self-regulation in the security products sector

There may be no British statutory regulation in the security products sector, but there are no shortage of standards, regulatory bodies, testing houses and trade associations. This sector also has some parts where there is comparatively strong regulation, due to the influence of the police and insurers, and other parts where there are no standards nor associations. This section will begin by discussing some of the many standards that are of relevance to security products and then analyse the regulatory arrangements for some of the different product markets.

Standards for security products

The most common and recognised standards in the UK are British Standards. They are published by the British Standards Institution (BSI) and written by all those with an interest in the standard, meeting to negotiate under the auspices of the BSI. With the advent of the single European market there has been a freeze on new British Standards for some products and services. There have also been initiatives to merge some existing national standards to create new Euro-wide standards or European Norms (ENs) as they are known. Some standards have also been further consolidated into international standards. Most of these are listed in Table 5.

Table 5. British Standards for the security products sector

Name of Standard	Description
	Intruder alarm products
BS 4166	Specification for automatic intruder alarm equipment terminating in police stations.
BS 4737	Intruder alarm systems in buildings.
	Part 1. Specification for installed systems with local audible and/or remote signalling.
	Part 2. Specification for installed systems for deliberate operation.
	Part 3. Specifications for components (various sections).
	Part 4. Section 1. Code of practice for planning and installation.
	Part 4. Section 2. Code of practice for maintenance and records.
	Part 4. Section 3. Code of practice for exterior alarm systems.
	Part 5. Recommendations for symbols for diagrams.
BS 5979	Code of practice for remote centres for alarm systems.
BS 6707	Specification for intruder alarm systems for consumer installation.
BS 6799	Code of practice for wire free intruder alarms.
BS 7042	Specification for high security systems in buildings.
BS 7150	Code of practice for intruder alarm systems with mains wiring communication.
	Integrated systems
BS 7807	Code of practice for combined and integrated systems.
	EAS equipment
BS 7230	Code of practice for article theft detection systems
	Personal attack alarms
BS 6800	Specification for home and personal security devices.
	Vehicle security products
BS 6803	Vehicle security alarm systems.
	Part 1. Specifications for installed systems.
	Part 3. Code of practice for the protection of vehicles and goods in transit.
BS AU 209	Vehicle security.
	Part 1a. Specifications for locking systems for passenger cars and car derived vehicles.
	Part 2. Specifications for security systems against theft in-car

equipment for entertainment and communication purposes.
Part 3. Specification for security marking of glazing for
passenger cars and car derived vehicles.
Part 5a. Specification for central power locking systems for
passenger cars and car derived vehicles.
Part 6. Specification for dead locking systems for passenger cars
and car derived vehicles.
Part 7. Specifications for locking systems for goods in vehicle
driver's compartments.

Locks

BS 455	Schedule of locks and latches in doors.
BS 2088	Performance test for thief resistant locks.
BS 3621	Specification for thief resistant locks.
BS 5872	Builders lock specification — locks which do not conform to BS 3621.

Security storage

BS 7558	Specifications for gun cabinets.
BS 7582	Code of practice for re-conditioning of used safes.

Fences

BS 1722	Specifications for fences.
	Part 1. Chain link fencing.
	Part 10. Specification for anti-intruder fences in chain link and welded mesh.
	Part 12. Specification for steel palisade fences.

Security glazing

BS 5051	Specification for glazing.
	Part 1. Bullet resistant glazing for internal use.
	Part 2. Bullet resistant glazing for external use.
BS 5357	Code of practice for the installation of security glazing.
BS 5544	Specification for anti-bandit glazing.
BS 6206	Specification for safety glazing.

Other relevant standards

BS 7480	Security seals.
BS 8220	Guide for security of buildings against crime.
	Part 1. Dwellings.
	Part 2. Offices and shops.

Note: At the time of writing some of the British Standards are being replaced by European Norms.

The BSI and its international superiors are not the only bodies that develop standards for
this sector. The Loss Prevention Certification Board (LPCB) keeps a directory of products
and services that meet British, European, international and Loss Prevention Standards (LPSs)

in the annual LPCB directory. In order for these products and services to be listed they are tested at the LPCB's laboratories at Borehamwood. These laboratories are the only ones accredited by the UKAS as independent in this sphere, and are the only British members of the European Fire and Security Group. The relevant LPSs for physical security products are also listed in Table 6.

Table 6. LPCB Loss Prevention Standards for security products

Name of Standard	Description
	Intruder alarm products
LPS 1167	Ultrasonic doppler detectors
LPS 1168	Microwave doppler detectors
LPS 1169	Passive infrared detectors
LPS 1188	Combined technology intruder alarm movement detectors
LPS 1200	Control and indicating equipment for intruder and hold up alarms.
	Safes
LPS 1183	Safe storage units.
	Physical security
LPS 1175	Security doors and shutters
LPS 1214	Physical protection devices for personnel computers
LPS 1224	Database management for asset marking

Source: LPCB (1999).

There are also a number of classifications for safes according to their standard of security, which range from Grade 0 to Grade V. These grades specify how much can be held in a safe for insurance to be given. For a Grade 0 it is £5,000, Grade I £10,000; Grade II £17,500; Grade III £35,000, Grade IV £60,000; and Grade V £100,000. This is another form of regulation because if an organisation wishes to hold £30,000 in a safe overnight, for insurance to be granted a Grade III safe would have to be used. At the bottom end of the market, where smaller amounts of valuables are stored, however, these grades are less important. A company wishing to gain the necessary grades must have its safe tested according to standards at one of the European testing houses, such as the LPCB's. Some trade associations and independent inspectorates have also produced their own standards and codes of practice.

Some security products are also tested and evaluated by a number of public bodies to assess if they are of a high enough standard to be used by organisations such as the police, prison service, customs and government departments. Many of these products are also used by the private sector, which makes these evaluations relevant because of the increasing movement of the private sector into traditional public roles.

One of the most prominent evaluating organisations is the Police Scientific Development Branch (PSDB). The PSDB's roles include providing technical advice to public bodies on equipment related issues and improving the operational effectiveness of the police, prison service and other clients. In undertaking its roles the PSDB has evaluated a range of electronic and physical security products. These products include CCTV cameras, electronic vehicle security products, explosives and weapons detection equipment, fences, intruder alarm detectors and surveillance equipment. Many of the evaluations of specific products are contained in what is called the 'Blue Book', which is a restricted publication (PSDB, 1995). Often a positive evaluation is essential before the government or other public body agrees to purchase the product. A positive evaluation is often marketed to the private sector. Indeed, some private organisations insist upon positive evaluations from the PSDB. Also based at the PSDB's laboratories is Sold Secure, which is an organisation dedicated to improving vehicle security. It develops specifications for security products for motor vehicles, motorbikes and cycles and tests products against these standards. Sold Secure has tested over 350 security products since its formation in 1992.

Another public body that tests and evaluates security products is the Special Services Group of the Security Facilities Division. The government's Defence Research Agency (DERA) also carries out research into some products that have a security purpose. DERA will also offer advice about other products including explosive and metal detection equipment, cameras and IT security, which are of interest to the industry. Assessments and lists of approved products are also produced by the security services for the government. There are also standards for some products, such as fencing established by the defence departments of different countries, which must be met in order to supply their armed forces.

The positive assessment of a body such as the PSDB is almost mandatory in most cases when a public organisation is purchasing specialist security equipment. For most private firms and some public bodies, however, the evaluation of equipment by these organisations is more a guide for purchasers. Nevertheless, a positive endorsement or the purchasing of these products by a government organisation often encourages the company to use the endorsement or buy in its marketing strategy.

The associations for manufacturers and distributors of security products

Manufacturers and distributors of security products join a number of trade associations, which often also set minimum standards for their members to meet. Some not only offer the traditional trade association services of promoting their sector, but also act as self-regulatory bodies. Most of the large manufacturers of security equipment are members of BSIA or some other association, which set minimum standards for membership. These include IPSA, the CCTV Manufacturers and Distributors Association (CCTVMDA), the Electronic Vehicle Security Association (EVSA), and the Electronic Article Surveillance Equipment Manufacturers Association, for example (see chapter 5 for full list of these associations). Given the large number of imports of these products into the UK, there are also the relevant associations of different countries to consider.

Regulation of the installation and operation of electronic security equipment

The voluntary regulatory arrangements covering the installation sector of electronic security products industry are far more complex. The strong involvement of the police and insurance industry in the intruder alarms sector also makes consideration of this sector on its own essential. The arrangements for the regulation of the intruder alarm installation sector are voluntary but, unlike other sectors of the industry, they are backed by strong sanctions from the police and insurance industry.

The aims of an intruder alarm are to deter and detect intruders by ultimately securing a police response to the place of alarm activation. Remote signalling alarms are linked either to the police or a central station. The police will respond once operatives monitoring alarms have assessed if it is a false alarm or not. Audible only alarms externally alert the immediate environment and aim to frighten off intruders and attract the attention of the public and ultimately the police. Therefore, persuading the police to respond to activations is essential to users of intruder alarms.

Responding to alarms places a huge burden upon police resources (ACPO, 1995a). The work alone of the police dealing with false alarms has been estimated as representing the annual employment of 6,500 full-time police officers (Drury and Bridges, 1995). These figures also exclude the effect of audible only alarms, which according to Fieldsend (1994) would roughly double the number of false alarms if they were also included. Given this substantial drain on police resources and the continued growth of alarm systems, it is not surprising the police have attached conditions to their response. These conditions are set out in the ACPO Intruder Alarms Policy, which was launched in May 1995 (the latest version is due to be published towards the end of 2000).

The national policy, subject to regional variations, sets conditions on police response to alarms. The policy sets out three levels of police response: level 1, immediate; level 2, desirable; and level 3, no police response. The level of response an intruder alarm receives depends upon the type of alarm, the circumstances of its installation and the status of the alarm. There are two types of alarm: type A, remote signalling; and type B, audible only. With the latter the police response requires additional factors, other than the alarm activating, such as human intervention (for example an individual seeing suspicious activities). The level of response to a type A alarm also varies according to the circumstances of the installation. The firms installing the alarms must be examined by a recognised 'independent' inspectorate to the relevant British Standard and linked to a central station meeting BS 5979 — although there are exemptions for some in-house central stations. An alarm not meeting these criteria may not gain a level 1 police response. The level of response can also change if there are continued false alarms. Following four false calls in a 12 month period, the level of response could be reduced to level 2. The policy also allows for those firms that are on the compliant list, regulated by a recognised independent inspectorate, to gain police vetting of their employees (ACPO, 1995b).

The conditions of the ACPO policy give very strong incentives for purchasers of intruder alarms to have them installed by firms that are regulated by recognised 'independent' inspectorates. It also gives these firms the incentive to subject themselves to voluntary regulation. The insurance industry adds an additional pressure by requiring many policy holders to have alarms installed by a firm recognised by an 'independent' inspectorate and by also offering discounts for installation by such firms. These two pressures combine to make it almost a necessity for an installation firm to be regulated by an independent inspectorates in the remote signalling market. Thus most installations are regulated by one of the independent inspectorates such as NACOSS (it must be noted that NACOSS also insists its installers meet BS EN ISO 9002 in addition to the relevant technical standards).

The installers of access control equipment, CCTV and integrated systems often overlap. Consequently, many of these installers are also registered with or are members of organisations dedicated to intruder alarm installation. However, the police and insurance pressure for intruder alarms installation is virtually non-existent for CCTV systems. Therefore, many installers of CCTV systems are not members of regulatory organisations or even registered with such bodies. The situation is exacerbated by the fact that there are no British, European or international standards for the installation of access control or CCTV systems, although there are NACOSS codes of practice.

There is also no regulation of the operation of CCTV systems in public places. Moreover, the last Conservative government removed the remaining limited regulations for the installation of CCTV systems with the CJPOA 94. This removed the need for planning permission for CCTV systems in public places (Davies, 1996). A number of other areas, including the placement of cameras, access to the control room, use of video footage, complaints procedures, and the conduct of operatives, are subject to no compulsory controls. Nevertheless, a number of voluntary codes of practice are in operation. The Home Office also has a model code of practice, which establishes guidelines on the control room, video tapes and still photography to name but a few (Home Office, 1994b). The Local Government Information Unit (LGIU) also recently developed a code of practice for the use of CCTV in public places (LGIU, 1996) as have the CCTV Users Group. At the time of writing the British Standard on the Management and Operation of CCTV Monitoring was also published (BS 7958).

The distinct nature of the electronic vehicle security sector is reflected by its voluntary regulatory arrangements. There is a trade association called the Mobile Electronics Safety Federation (MESF) and an independent inspectorate called the Vehicle Security Inspection Board (VSIB). In the EAS equipment sector EASMA sets standards of installation and operation for member companies.

Regulation of the installers of physical security equipment

The voluntary regulatory structure for locksmiths is undertaken by the MLA and its divisions. All members of the association are subject to a strict code of ethics, which covers integrity and honesty, abiding by the MLA's rules, impartiality, professional conduct and the promotion of the profession (MLA, 1994). For those organisations involved in the installation of fences there are two main bodies: the Fencing Contractors Association, and the National Fencing Training Authority (NFTA), which is the equivalent of SITO for the fencing industry. For the installation of glazing there is a degree of statutory intervention, as building regulations issued under Health and Safety legislation often mandate the installation of a specific type of glass to certain standards set by the appropriate British Standard.

Conclusion

Although it can be broadly split into electronic and physical parts, the security products sector is probably the most fragmented sector of the private security industry. This chapter has demonstrated the very wide range of security products that exist. It has also discussed the very different characteristics of the markets that produce, distribute, install and maintain them. The very complex and varied regulatory structure that governs this sector has also been examined. The future of the security products sector poses a number of challenges. As security continues to be in demand, so will security products, and as the advance of technology proceeds, so will their sophistication. Many will not only be used by security personnel to aid their roles, but also to replace them. At one level these changes will increase fragmentation of the industry with new products emerging; but at another level many single products will replace a multiplicity of products from varying sectors, thus providing an integrating force. All of these processes will pose new challenges for the setting and enforcement of standards.

Chapter 11

The Margins of the Private Security Industry

Previous chapters explored the main services and products of the private security industry. They also showed the difficulties in drawing a line where the private security industry begins and ends. In this chapter we shall explore this indistinct zone or what we have called the 'margins' between the private security industry and other industries and policing activities. In chapter 2, we considered the problems of defining private security. In doing so we developed a number of tests that could be used to consider, amongst other things, the degree of 'privateness' and 'securityness'. In this chapter some of the activities that are not unambiguously private security will be considered. One of the most common dichotomies is between the public police and private security. This chapter will start by exploring the large grey area between these. It will then move on to review some of the many services and products provided by companies, which are not unambiguously private security. Finally, the chapter will consider some of the many occupations that have a significant security element, but are not generally considered as private security, as well as some of the many voluntary roles with a security function. This chapter is intended to give a representative sample of some of the many occupations and organisations that fit into this domain, as there are potentially many more. Therefore, the following should not be considered exhaustive.

The margins of private security and the police

There is no simple line dividing the police (by which we are referring to Home Office police forces) from private security. There is a huge complex web of both private and public organisations engaged in policing, which are neither unambiguously the police nor private security. Many of these undertake functions that in varying degrees can also be found in the private security industry. This web is made up of 'specialised police forces, such as the British Transport Police; state enforcement and investigatory bodies, such as Benefits Agency investigators; local authority enforcement and investigatory bodies, such as trading standards officers; and private policing investigatory and enforcement bodies, such as the RSPCA. The wide range of bodies that fit this category vary in the degree to which they resemble a police or private security organisation. Johnston (1992a) has described these phenomena as 'hybrid' policing bodies, whereas Jones and Newburn (1998) have called them 'other bodies of constables' and 'other public policing bodies'. The different organisations that fall into this grey area will now be considered.

Policing bodies with constabulary powers

When most people come into contact with police officers it is usually one of the 52 constabularies of the United Kingdom. These are not the only police forces, however. There are a number of special police forces that operate in specialist areas, have uniforms almost identical to the main police forces, and hold the same constabulary powers (although usually restricted to a limited geographical/functional area) and are commonly called non-Home Office police forces. There are also other state uniformed organisations who hold similar powers, but which are separate from the police — for example prison officers, customs officers and water bailiffs.

Some of these non-Home Office forces, because of privatisation, now serve private companies. The Port of London Police serves the Port of Tilbury, which is now a private company. Others serve a mixture of public bodies and private companies such as the United Kingdom Atomic Energy Authority Constabulary (UKAEAC), which polices both privatised nuclear facilities and some of the publicly owned facilities. Hence some private companies could be described as having their own private police force. Although it could be argued the office of constable, with the duty to be independent, would not be compromised by any private pressures on their role, it seems difficult to believe that those constables, who depend upon the private companies funding and support, would not be influenced, as ultimately 'he who pays the piper calls the tune'. Some of these forces also serve only public bodies, but could be argued to be little more than 'glamorous' in-house security departments with special powers. The Royal Botanical Gardens Police, for instance, provide a policing/security role for the Royal Botanical Gardens. The UKAEAC is another body that has been described as little more than a very expensive in-house security department. Indeed, when the very low and different levels of crime are considered it does seem atypical. For instance Johnston (1992a) found the UKAEAC force had only made 53 arrests between 1977 and 1992.

There are also non-uniformed examples with constabulary powers, such as the water bailiffs of the Environment Agency, who possess the powers of the constable under the *Salmon and Freshwater Fisheries Act* 1975. The roles that some of these forces undertake are often comparable to some private security departments for large organisations. They also undertake crime prevention, order maintenance, loss prevention and protection functions. In terms of the public/private test, however, the officers employed hold special powers and, as 'sworn constables', are duty bound to serve the 'public interest'. For these reasons they are more appropriately placed at the 'margins' of the private security industry.

Many local authorities establish their own police or security forces, which do not have the powers of a constable, but which have special powers to enforce local bye-laws. Similarly some transport organisations, such as the various rail companies, the London Underground and bus companies, have revenue protection officers to ensure that those using their services have valid tickets. Some of these have special powers under bye-laws. Similarly, many local authorities have traffic wardens to enforce parking regulations. Again some of these occupations could be viewed as both special policing bodies and as in-house security departments in terms of the security functions undertaken, particularly loss prevention. The possession of special powers makes their inclusion in the 'margins' section the most suitable.

State enforcement and investigatory bodies

There are a multiplicity of public organisations that carry out enforcement and investigatory functions that employing officers. In doing so, many undertake roles that are also performed by the private security industry and, in varying degrees, pursue crime prevention, order maintenance, loss prevention and protection functions. The following represent a sample of some of the organisations that fall into this category. The Vehicle Inspectorate is responsible for enforcing road safety legislation, monitoring the MOT scheme, issueing of MOT certificate pads and the statutory testing of heavy goods vehicles and public service vehicles. Some of its investigatory roles are comparable to those of private investigators. They are also engaged in crime and loss prevention functions. However, they serve the public interest and possess special statutory powers, hence their inclusion within the 'margins'. A similar case can be applied to the Radio Communications Agency, which has inspectors to carry out a range of functions relating to the enforcement of broadcasting legislation such as detecting illegal radio transmitters. Other government organisations are involved in activities that are open to abuse and consequently employ investigators to ensure that their resources are being used properly. The Benefits Agency is responsible for paying social security benefits and is subject to fraud by some claimants. To combat this, the agency employs a force of 3,000 investigators to detect fraud. It even has its own prosecutions department to bring prosecutions where necessary. Despite their comparable roles to some private investigators and their pursuit of crime prevention and loss prevention functions, the limited additional powers they possess place them, in our opinion, at the 'margins'. We do accept, however, that a strong case could be made for their inclusion within the private security industry. There are many other state regulatory and enforcement agencies that a perusal through the *Civil Service Year Book* that reveals can be located at the 'margins'. Most are clearly 'policing' bodies, but some could also be compared to in-house security departments. Therefore most of these bodies occupy the grey area at the 'margins' between the private security industry and the police.

Local authority enforcement and investigatory bodies

Local authorities have a wide range of statutory regulatory functions to perform. To undertake these roles local authorities employ investigative and enforcement officers. These include Trading Standards Officers, who enforce the law on matters such as counterfeit goods; and Environmental Health Inspectors, who investigate matters related to environmental health. Local authorities are also responsible for collecting monies such as the council tax and distributing benefits, such as housing benefit. To prevent fraud in these areas investigative officers are also employed. An example of this type of officer is a Housing Benefit Detection Officer. The *Crime and Disorder Act* 1998 has led to a huge increase in local authorities employing community safety officers, who also pursue strategies aimed at preventing crime and enforcing order in the community. As with state enforcement and investigatory bodies many of these activities also have private equivalents and effectively have crime and loss prevention, and protection functions. Many also serve the public interest and/or possess special powers making their location in the 'margins' more appropriate.

Private policing investigatory and enforcement organisations

There are also some organisations that are private and have no special powers, but act as quasi-police forces. The most prominent of these are the RSPCA, the Royal Society for the Protection of Birds (RSPB) and the National Society for the Prevention of Cruelty to Children (NSPCC). These have been placed at the 'margins' because, although they are private organisations engaged in crime prevention, order maintenance and protection functions in varying degrees, they serve the public and not a private interest. These organisations are open to any member of the public who wishes to make a complaint or to use their services. For these reasons, combined with their unique roles, we felt they could not be clearly placed within the bounds of the private security industry.

Services and products on the margins of the private security industry

In chapter 2, when we attempted a definition of the private security industry, we noted there were a number of services and products where a significant element of the service or product's role was security related. It is important to stress yet again, however, that during the research for this book there were some who argued that some of these products were 'security' as well as others who argued they were not. After assessing them we considered they fit into the 'margins' of the industry. The test we have used to distinguish whether they are on the 'margins', is the extent to which the four functions discussed in chapter 2 can be applied to them and whether the security activity constitutes the primary role of the product or service. Other factors have also been identified where relevant. The following list of services and products is also not exhaustive, as there are others that could fall into this area, but which we have not had time to consider.

In chapter 6, the CIT sector was considered and it was illustrated that couriers undertake similar activities and that some would also regard them as a part of the private security industry. After an analysis of what they undertake, however, they were more appropriately placed at the margins. They are often regarded as part of the private security industry because of their transportation of some valuables and sensitive documents. The fact that some security companies, such as Group 4, are also involved in these activities adds to this. If couriers were undertaking this function the majority of the time, there would be a stronger case for them being part of the private security industry, rather than on the 'margins'. Most of the time, however, couriers are not transporting valuables or highly sensitive documents. The vehicles that are used generally have no special security measures used to protect what is carried, and generally most couriers do not vet their staff to the same standards as a security company would. In terms of the four functions of private security, most of the time a courier is not engaged to transport documents as a crime or loss prevention measure. They are clearly not engaged in order maintenance, although they do have a protective role. All this evidence leads us to place couriers on the 'margins' of the private security industry.

There are a number of employment agencies that offer services to the private security industry. The services they offer generally fall into two categories: the recruitment of permanent security employees for an organisation and the supply of temporary security personnel. In some ways employment agencies are supplying the same service as contract security companies by supplying security personnel. Therefore, if the agency's business was largely security related, it could be classed as a part of the private security industry. However, it is those agencies that offer their services to a range of clients, of which the private security industry is a minority, that cause problems in classification. In these cases they could be regarded on the margins of the industry.

Some of the other services that could be considered on the margins of the industry include credit check agencies. They primarily undertake investigations into the financial status of a prospective borrower, but their investigations can also prevent fraud. Such activities are crime and loss prevention related functions, which brings them in to the 'margins' of the private security industry. Many companies are now starting to test their employees for drug abuse. This may involve an in-house kit, or the taking of samples and sending them to a specialist laboratory. The ultimate aim is to prevent drug abusers gaining employment or continuing in employment, as the effects of their drug abuse might affect their performance at work or even lead them to commit crimes against the organisation. In this capacity such testers could be viewed on the 'margins' of the industry because of their loss and crime prevention functions.

Occupations with significant security activity

There are a number of occupations where security forms a significant part of the duties, but not at such a level as to warrant being termed a security occupation. Some of the occupations that fall into this category include caretakers, car park attendants, cleaners, facilities managers, gamekeepers, porters, receptionists and risk managers, to name the most common. Some of the other occupations that we do not have space to consider include: investigative journalists, museum curators, park rangers, and computer operating staff . The following analysis will give a general overview of these occupations. Particular examples, however, might have more or less security duties. Some of the occupations we have considered as part of the industry might be regarded by some as on the margins. For example, during the many interviews conducted for this research some regarded stewards as not private security and some did not even regard door supervisors as such. However, in our analysis we decided they warranted inclusion within the bounds of private security because of the significant security duties they undertake (see chapter 2 and 6). It is also of note that Jones and Newburn (1998), in their analysis of the private security industry, offer some evidence that roles such as caretakers, porters, bus conductors etc have been replaced by private security personnel. This illustrates that the security function has become so important in some jobs that they have become pure security roles. The following brief discussion will illustrate some of the occupations on the 'margins' of private security industry because of the significant security aspects in their roles. Again this list is not meant to be exhaustive and there may also be other occupations, which we have not considered, that could also fit into this classification.

Caretakers in colleges, schools, universities and other organisations such as local authorities undertake a wide range of functions. These usually include mainly cleaning and maintenance. Many are also responsible for the locking and opening of premises, which often involves the setting of alarms. They may also be involved in the patrolling of buildings before they are closed to ensure they are secure. This might include ensuring all windows and doors are locked and there is no one present in the building. Their presence is often also advocated as a form of crime prevention to add to the scope of surveillance (Gilling, 1997).

Car park attendants' duties frequently include security functions such as access control, transporting cash from ticket meters and surveillance. For most carpark attendants these would not be the main duties, but they do illustrate the security element. Most cleaners would not regard any of their duties as security related. Some, however, have instructions to be on the look out for suspicious packages. In some locations, because of their familiarity with the location they clean, they are the first involved in a search, if there is a bomb threat, as they are the most likely to identify anything suspicious or unusual.

Facilities managers have responsibility for a wide range of functions related to the management of property. These can include catering, cleaning, environment management (air conditioning, plants etc), health and safety, maintenance and of course security. Often when facilities managers have responsibility for a major shopping centre this security function can be a significant part of their duties.

Gamekeepers and their sister occupations stalkers (deer) and gillies (game fish) undertake a wide range of duties, many of which could be described as security related. Indeed a strong case could be made for them being a form of rural in-house security, an armed force as well! Some of their security duties include the prevention and detection of poaching and other countryside crimes and related problems. As a significant part of their duties involve the maintenance of the environment to secure the survival of the game they care for, it was felt they could not be identified clearly as security personnel.

Hospital porters may also have significant security duties. In a survey we carried out about in-house security many answered the questions with regard to security for their portering staff (Button and George, 1994). One of the more common duties they undertook was providing a security presence in the accident and emergency department, where there are often public order problems. In other locations, such as local authorities and educational establishments, porters have security roles such as setting alarms, closing and unlocking doors, patrolling corridors and buildings. Some porters often live on site and have a quasi-resident security role. These types of security roles are similar to those undertaken by some caretakers in schools.

Many receptionists undertake a security role as part of their duties. Moreover, some security officers undertake reception duties. The security roles undertaken by receptionists may include access control functions such as checking passes and signing in visitors. They may also have a role such as monitoring a CCTV screen, amongst their non-security duties. Risk managers have responsibility for managing all the risks to an organisation. These risks usually cover disasters, health and safety issues, investments, insurance and of course security, amongst others. The degree of security risks varies across different organisations and thus so does the extent of interest in security if a risk manager is employed. In some organisations, such as aviation and retail security, risks are very significant.

Voluntary security roles

There are also a wide range of voluntary roles that have a significant security element. At one end of the spectrum are the many individuals who volunteer their spare time to become special constables. Less formally organised roles include the many people who take an active interest in neighbourhood watch. For those who are very keen and who spend time undertaking surveillance and reporting to the police suspicious activities, they are almost a new form of the 'watch'. The National Trust also uses several thousand volunteers, who undertake security and stewarding functions in their properties. Many organisations that organise demonstrations and protests usually organise their own stewards, who often have many roles related to ensuring protesters keep within the law and particular rules of the demonstration and organisation. These are just a few examples of the many voluntary roles that involve a significant security element.

Conclusion

Chapters 7 to 10 considered the different sectors of the private security industry. This chapter has identified what we have termed the 'margins' of the private security industry, which contains a multiplicity of organisations, occupations, services and products that do not clearly meet all the criteria which we argue distinguish private security, as discussed in chapter 2. Such is the nature of the industry and policing bodies it is difficult — if not impossible — to draw the line dividing the boundaries. This zone, which we have called the 'margins', we hope goes at least some of the way to addressing this.

Part 3. An Integrated Service:
Case Studies in the Use
of Private Security

Chapter 12

The Use of Private Security:
An Integrated Service

Private security is an omnipresent phenomenon that affects almost everyone. An individual locking the front door when leaving home, walking past security officers when entering work, or tapping in a personal identification number when withdrawing money from a cash point, is dependent on the services offered by the private security industry. The extent, type and use of private security by organisations (and individuals in domestic circumstances) varies considerably between different sectors of industry and commerce. It is therefore the purpose of this section of the book to explore how different organisations make use of private security, the diverse security risks they face and the origin of these risks.

This chapter examines the decision-making process for employing and purchasing private security. The following chapters will then explore four case studies of security in the retail, aviation and entertainment sectors, as well as the Ministry of Defence. These studies will illustrate how security risks, their origins and private security strategies vary. It will also show that although many different and distinct sectors offer security services and products, when they are used by an organisation it as an integrated and strategic service.

Why private security?

The *raison d'être* for the use of private security by most organisations and individuals is protection against crime. This is not the sole purpose as it may - depending upon the particular circumstances — also be used for protection against other risks such as fire and natural disasters, as well for order maintenance and the enforcement of internal rules. Generally, private security is used to protect four areas. The first area involves the protection of people. This may include the employees and customers of an organisation or the general public. The second area involves protection of an individual's or organisation's physical assets, for example: a home, building, vehicle or capital equipment. The third area of protection is the information of an organisation. Many establishments have sensitive information, such as product specifications, financial plans and marketing strategies, which could be very useful to a competitor. The final area of protection is the reputation of an individual or organisation. Reputation is very important and the emergence of a negative image could result in serious damage in some cases. Take for instance a food manufacturer who has some products contaminated by a blackmailer; the consequent negative publicity could lead to a massive reduction in sales. Therefore, private security is used, amongst other strategies, to protect people, physical assets, information and the reputation of an individual or organisation against the threat of crime and related problems.

The decision-maker for private security

The individual who makes the decision on which security strategies to use varies immensely. Broadly there are three types of people with responsibility for security. First, there are ordinary citizens who purchase security products and services to protect themselves and their property. Second, in some organisations managers or other personnel have a range of responsibilities, one of which is security, and thus have a role in developing security strategies. The final type of individual with responsibility is a specialised security operative. As the vast majority of the security market is concentrated in the commercial sector, the discussion will focus on the latter two types of decision-maker.

Generalists

In the overwhelming majority of organisations the individual responsible for security is a generalist rather than a specialist. In most cases this is because the organisation is not large enough or the risks do not warrant the employment of a specialist security operative. In many small firms the chief executive or a senior manager will take responsibility for security amongst other duties. In the public sector there are also many generalists with responsibility for security. For example, head teachers have responsibility for security in many schools and general managers often assume a similar role in hospitals. Many public sector organisations have such limited budgets that they cannot afford to employ a security manager, although some with enough resources are now starting to hire specialists. Some large commercial organisations with a huge turnover also do not employ a specialist security manager. In the case of some water companies the perceived minimal security risks do not warrant the employment of a specialist, as water is plentiful and cheap and not worth stealing.

Other organisations employ personnel with responsibility for a range of specialised duties, including those with an expertise in security. Some of the occupations include facilities, personnel, contingency planning, risk and safety managers. In some organisations these occupations may retain overall responsibility for security even when specialists are employed (see adviser/manager dichotomy later in this chapter). Facilities managers are usually responsible for an entire building such as an office block or shopping centre. They are responsible for a range of cleaning, maintenance and refuse collection services as well as security. A personnel manager might be responsible for a range of functions such as industrial relations, recruitment, training, safety and security. In many small firms the personnel manager is responsible for hiring a contract security firm or purchasing security products. Risk managers often have responsibility for a wide range of activities, which may include health and safety, investment appraisal, insurance, corporate strategy as well as security risks (Borodzicz, 1996). Finally, safety managers in some organisations may also have a responsibility for security in addition to their general health and safety functions.

Many of the managers with responsibility for security will seek help from experts by turning to the local crime prevention officer, an independent security consultant or a security firm. Even some security specialists may seek external assistance. In some cases they require specialist expertise on a particular subject but in others they may simply have a policy of seeking external sources of security advice. Hence, such specialists are often important in the decision-making process. For instance, a school may decide it wants to improve security by installing CCTV but the head teacher has no idea of how many or what type of CCTV cameras to purchase. The head teacher may therefore turn to a security consultant who draws up a specification for firms to bid for. The consultant may even be invited to participate in the awarding process for the contract.

Security specialists

Many organisations employ security specialists to undertake a range of security roles. They have a multitude of different job titles. Some of the most popular include: security manager, security adviser, security controller, security co-ordinator, security administrator, chief security officer, security director, security analyst, security specialist, loss prevention manager, director of surveillance, and protection manager. They may also be preceded by a range of titles such as assistant, group, senior, divisional and deputy. Security specialists have a wide range of duties and some will have responsibility for non-security related functions such as health and safety and fire prevention. Differences are also evident in relation to whether they report to a senior director or some other manager below board level. Research by Hearnden (1995) found that 58 per cent of those security managers surveyed reported to a director or main board. Some security specialists also have no responsibility for certain security functions. For instance, information technology security is often hived off to a specialist in the IT department. In some organisations crisis management is the responsibility of a specialist crisis manager, contingency planner or some other manager. A number of organisations may also have a separate investigations department.

Some security managers also have additional non-security related functions. Hearnden (1995) found the most common additional function undertaken by security managers was fire prevention and response. In earlier research he identified training, customer liaison, safety and recruitment as other common responsibilities (Hearnden, 1993). Perhaps the main difference between specialists is the extent to which they have a responsibility for the management of security, or whether they advise general managers on the use of security. In most instances the security specialist may have a mixture of responsibility somewhere between the two.

Despite these necessary qualifications it is possible to identify two models of the security specialist that could be broadly applied to most security roles. The first can be termed the security adviser and the second the security manager. The responsibilities of the former include: developing security policies; providing advice to managers on the risks faced and the most appropriate strategies to combat them; advising on the selection and monitoring of security contractors; monitoring incidents and losses; and identifying trends. The latter has responsibility for the management of security. This may encompass the assessment of risks, the selection of appropriate security strategies, the management of security staff, the management of a crisis and, ultimately, the control of a security budget. The nature of most security managers' functions means that they may also assume similar roles to security advisers. Essentially the main difference is that of responsibility, with the security manager having the ultimate authority over security. In the case of a security adviser the line or other managers assume this responsibility. The reality of the commercial world means that there are many security specialists who undertake both advisory and managerial roles in varying proportions and in some organisations there might be both security advisers and managers. The following case studies illustrate the two roles and the complex issues that arise when comparisons are made.

Case study 1 involves a major financial services company with a multibillion-pound turnover that employs many thousands of staff and has several million customers. The company has a security department, with four employees and an administrator, which is headed by a Group Security Adviser (GSA). The GSA's role amounts to developing security policy, helping to identify risks and the best security strategies, and assisting with the selection and monitoring of security contractors. Line managers are ultimately responsible for security and any security staff required, are contracted in by them. However, the GSA takes a direct role in investigations of fraud, counter-intelligence and crisis management — such as a blackmail attempt or the kidnapping of a senior executive.

Case study 2 involves a large food manufacturer with a multimillion-pound turnover and several thousand employees at three manufacturing sites. The company has one Group Security Manager (GSM) supported by a team of nearly 50 in-house security officers and occasional contract security officers. The GSM reports to the personnel director and has responsibility for the security of all three sites, except computer security, and has a budget to achieve those aims. The GSM has responsibility for assessing the risks and then employing the appropriate security strategy to address them. The GSM is directly in control of all in-house and contract officers, with an additional responsibility for any investigations into frauds.

Case study 3 involves a major leisure company with a multimillion-pound turnover and several thousand employees. The company's portfolio includes nightclubs, leisure facilities and holiday camps. It employs one security adviser whose responsibilities include development of security policy, advice on security procurement, training of staff in security and special investigations. The decisions on security at a nightclub would be the responsibility of the nightclub manager, subject to the company security policy and after consultation with the security adviser.

Case study 4 involves a major vehicle manufacturer with a multibillion-pound turnover and several thousand employees. The company has a number of manufacturing sites over a very large area. The Company Protection Services department is headed by a director with a staff of nearly 300. In addition to security, the department is also responsible for fire prevention, fire fighting, ambulance aid and investigations. The in-house force also offers itself for hire to outside contractors. The director of this department therefore not only manages security but also other significant functions.

It would also be useful to briefly describe the general background of security specialists, as there is a common perception that they are former police officers or ex-servicemen. According to research carried out by Hearnden (1993) this is accurate, but the percentage is declining. In 1989, no fewer than 86 per cent of security managers were recruited from a military or police background. In 1991, this had declined to 76 per cent and by 1993, it was 61 per cent. Given that security management is often a second career, it is no surprise to discover the average age found by Hearnden (1995) was 50.2 years. Almost three security managers in five left school at the age of 17 but on average they achieved higher than the national average at GCE 'O' level and about average at the GCE 'A' level. He also found security managers worked on average 50 hours per week and earned on average £25,796 per annum (Hearnden, 1993). More recent research by Jones and Newburn (1998) found that 17 per cent of firms in the security industry employed former police officers, although this covered all ranks and not just the more senior personnel.

The decision-making process

Whoever is responsible for security, whether it is an ordinary citizen, general manager or specialist, decisions need to be made on the strategies to address certain risks. Most will undertake the following decision-making process in varying degrees of sophistication to achieve those aims. First, a risk assessment will be undertaken, then options will be identified to address those risks and finally the risks will be continually managed. The most basic risk assessment might involve an ordinary citizen who perceives that the risk of burglary has increased after a neighbour has been burgled and decides to purchase an intruder alarm. A more sophisticated risk analysis might involve a security manager undertaking a detailed survey of a company and analysing statistics over a period of time. It also important to note that in larger organisations the decision-making process may take place at national, regional and local levels.

A risk assessment, also known as a security risk assessment and security survey, has been described as:

> a qualitative and quantitative process of prioritising an organisation's security risks by their likelihood to occur, their impact to the organisation, and their potential cause to harm persons (Somerson, 1996: 2).

The process involves identifying security risks to an organisation, which may also include other non-security risks, determining how often they occur or are likely to occur and their potential consequences. These risks could include arson, burglary, fraud, theft and robbery. The risks confronting the various sectors of the industry are extremely diverse. For instance, the main risks addressed by security operatives in the aviation industry are the threat of bombs and hijacking by terrorists. These events occur very rarely but, when they do, the consequences are extremely serious. For instance, the bombing of the Pan Am Flight 103 over Lockerbie was a significant contributor to the airline going out of business. In retail security such risks are virtually non-existent but the main threats, shoplifting, staff theft and fraud, are much more frequent. If left unchecked, these problems could be just as damaging to the financial viability of an organisation.

Once potential threats have been identified in the risk assessment, strategies to address them must be pursued. There are a number of options open to the decision-maker. The first option could be no action at all. It might be decided that the risk is minimal and even if it occurred the damage would be negligible. In such cases a decision-maker would conclude that the cost of addressing a potential risk is not economically viable. For instance, the owner of a pub with a large car park might decide security measures, such as CCTV, were unnecessary if no cars had been stolen in the past. Similarly, a shop may decide that counterfeit notes are so rare that purchasing specialist equipment to detect them is unnecessary.

A second strategy that could be pursued by an organisation is to avoid the risk altogether. Thus a company that has suffered repeated robberies in the evening might decide to relocate. A takeaway food firm might decide to cease deliveries to a particular area after a succession of bogus calls, harassment and assault of staff delivering the food.

Another strategy frequently employed is transferring or contracting out the risks to someone else. For example, many banks employ CIT companies to carry their valuables and assume responsibility for them. A company experiencing losses in its catering department through theft and fraud, which cannot be addressed, might decide to contract out the whole division.

The use of insurance or self-insurance is another strategy that could be pursued by an organisation. Thus in the event of the risk occurring the organisation will at least be able to cover a significant amount of the costs by compensation from the insurance company. In the case of self-insurance the company holds financial reserves to deal with costs in the event of the risk occurring. Insurance can, however, be expensive and the insurance company may require an organisation to purchase additional protection. Ultimately, the insurance company might not even pay out or take a long time to do so.

The most common strategy for dealing with risks is to seek to reduce the likelihood of them occurring. A wide number of security strategies could be pursued including the employment of security officers and investigators, the installation of intruder alarms and other security equipment, and the introduction of security procedures. Hence, a factory that fears a burglary at night may purchase an intruder alarm linked to a central station. Many celebrities and public figures risk being harassed, attacked or assassinated, and so close protection officers are employed to address this threat. In hospitals the risk of babies being abducted from maternity units has increased in recent years and the introduction of tagging systems is one of the security strategies implemented to address this problem. The increased risk of intruders entering some schools has led some head teachers to employ security officers. These are just a few examples of how security risks can be reduced, a subject that will be explored in greater depth in the case studies in the following chapters.

Once the individual responsible for security has undertaken the risk assessment and decided which strategy to employ, it is then a question of managing the risks. Thus the risks and the effectiveness of the strategies must be continually reviewed and the strategies adapted to changes in the nature of risks.

In some instances the nature of the risks and/or the performance of organisations in combating them has led to government intervention. With some risks, notably terrorism, it is not possible for the risk assessor to accurately assess the threat. Therefore, in UK aviation security, for example, the regulatory body TRANSEC disseminates information on the potential threat posed by terrorism. In other areas of security the addressing of certain risks has been ineffective and the government or some other body has intervened. The extent of intervention has ranged from the development of statutory standards on what security measures should be implemented, to voluntary guidelines or codes of practice. It may also be necessary in some cases for government to amend legislation, creating new offences and/ or tougher penalties to help deter certain behaviour.

The characteristics of security in different sectors

Clearly the nature of the security response is dependent on the perceived level of risk. In different sectors the risks vary immensely and this is reflected in the diverse security strategies that are employed to combat them. The following figure illustrates how decision-makers are influenced in their final judgements.

Figure 4. A model of the security decision-making process

> Nature, origin and consequences of risks
> Any statutory/voluntary requirements
> Strategies to address risks

When the decision-maker is considering different strategies to address a risk there are usually a wide range of options that could be pursued. Take for instance the risk of shoplifting in a large supermarket. The first decision is likely to involve some form of risk reduction. Once this has been agreed a wide range of strategies are available to the decision-maker. Any one of the following examples could be pursued either separately or in combination with other policies: the use of security officers, store detectives, electronic article surveillance systems, loop alarms, mirrors and CCTV systems. Thus the decision-makers are looking at products and services from the whole spectrum of the private security industry. The use of all the products is designed as an integrated service for the prevention of shop theft. Nevertheless, they may also have the objective of addressing other security risks and even non-security risks. The following chapters 13 to 16 will illustrate these differences in four distinct case studies.

Chapter 13

Retail Security

Retailers face significant challenges to their profits from the criminal fraternity, ranging from shoplifting, fraud and robbery to terrorist attack. It has been estimated by the British Retail Consortium (BRC) that the total cost of crime and crime prevention to retailers was in the region of £1.9 billion during 1997-98 (BRC, 1999). Given these costs it is not surprising that numerous security initiatives have emerged to address the many different security risks to retailers' profits. Numerous studies have evaluated the cost of these problems and the many different security strategies employed to combat them. Therefore, retail security is one of the few areas of the industry with a wide range of literature to draw upon in comparison to the dearth of research in many other sectors (Beck and Willis, 1995 and Jones, 1997). In this chapter we will explore the many problems faced by retailers, both independently and collectively, in a shopping and/or town centre. This chapter will also explore some of the many security strategies that retailers have developed individually and collectively. Before we embark upon this, however, it would be useful to describe what we mean by retailing. For the purposes of this chapter retailing will be used to describe a wide variety of 'high street users' as well as those that have moved to out-of-town shopping centres and are covered by the BRC. These categories include: booksellers; chemists; clothing; DIY; hardware; china; electrical; gas; electrical hire and repair; footwear and leather goods; furniture, textiles and carpets; grocery retailers; off-licences; and food stores (BRC, 1998).

The range of security risks and problems facing retailing

The range of threats and problems facing retailers can be divided into two categories. The first category encompasses direct threats against the retail unit such as shoplifting in the store. In contrast, the second category covers those problems faced by the retail community as a whole, which indirectly or directly affect the retail unit.

Theft by customer

Theft by customer, or shoplifting as it is more commonly known, accounted for the largest proportion of the £1.39 billion cost of crime estimated by the BRC during 1998. Shoplifting accounted for £604 million, or 43 per cent, of the cost of crime to retailers. The survey also found that retailers apprehended nearly 800,000 suspects for shoplifting of which just under 500,000 were handed over to the police. Shoplifting represents an enormous problem to retailers, but only certain retailers. For example, shoplifting is not a problem to those in the retail community who only offer services. Neither is it a cause for concern for retailers selling large goods that are not easily hidden or carried out of a store such as furniture or a carpet. Theft by customers is similarly not a concern for retailers whose goods are behind the counters or in secure displays, such as jewellers. Shoplifting is, however, a major risk in stores with open displays of goods, which can be easily hidden and taken away, such as supermarkets, clothes stores and bookshops.

Theft by staff

One of the biggest risks to any retailer is dishonest staff using a privileged position to steal from their employer. This is often described as 'shrinkage'. In common with shoplifting a variety of methods are used to commit staff theft. The more frequently documented examples include giving away stock to customers. A supermarket cashier might, for example, only charge for a fraction of the goods chosen by a customer. Another practice is the cashier charging a customer the full price but only ringing up a fraction of it and pocketing the difference. Other practices may include short-changing the customer (giving change for a £10 note when a £20 note was given, for example). Staff theft at its simplest may just involve filling up a bag with goods before going home or passing on a bag full of stolen goods to a customer. There are simply many ways for staff to cheat their employer (see, Mars, 1982; and Gill, K. 1994). The extent of staff theft is by definition very difficult to assess because one cannot be certain what has caused losses. It is just as likely to be customer theft or poor accounting. Nevertheless, the 1998 Retail Crime Survey (RCS) estimated the total cost of this crime to retailers to be in the region of £364 million, constituting 26 per cent of the total cost of crime (BRC, 1999).

Burglary

The risk of burglary to retailers has been transformed in recent years with the growing phenomenon of the 'ram raid'. A typical example involves a stolen car reversing into the front window of a store and raiders stealing all the goods in the vicinity of the window. The 'ram raid', however, is not new. Its origins can be traced to the beginning of the mass use of the motor car in the period before the Second World War (Jacques, 1994). It is only the high profile 'ram raiding' has been given by the media in recent years that has created the image of a new crime. Since no special statistics are kept by the police ('ram raiding' is included with all other offences of commercial burglary), it is difficult to assess the extent of this type of crime. Other types of burglary include the 'typical' breaking and entering, climbing into a shop through an open or broken window, forcing a door or even knocking down a wall from a neighbouring store to gain entry. According to the 1998 RCS, retail premises were the target of over 78,000 burglaries amounting to gross losses of £110 million.

Criminal damage

Criminal damage ranges from the smashing of windows, to the writing of graffiti and destruction or vandalism of a retailer's property. In a survey of Asian-run small shops, 33 per cent had a window smashed, 10 per cent had graffiti written on their building and 6 per cent had experienced some other form of criminal damage (Ekblom, Simon & Birdi, 1988). Some offenders may also intentionally damage the goods in a store during an attempted theft or for some other purpose. For example, one of the authors, while working as a security officer in a retail store, witnessed a man committing a sexual act on swimwear, which was for sale. The motives for criminal damage range from attempted burglaries, the activities of delinquent youngsters, drunks who feel like smashing a window, to sexual deviants. According to the BRC, retailers experienced 77,000 incidents of criminal damage during 1998, which amounted to gross crime costs of £26 million.

Arson

Arson is one of the rarer risks for retailers but when it does occur the cost can be considerable. The average incident of criminal damage costs retailers £337, but the average act of arson costs £2,000. During 1998 there were 4,500 arson attacks costing a total of £9 million (BRC, 1999). Arson is often committed by burglars who set fire to the premises during the process of a raid. In other cases it is the work of disturbed individuals who simply want to set fire to a building, or extremists who commit racist attacks on small shops owned by ethnic minorities. Attempted insurance fraud is another frequent explanation.

Robbery and till snatches

Robbery can be undertaken by organised gangs using firearms and pursuing their crime with military precision, or it can be conducted by a lone individual wielding a knife, desperate for money to buy his or her next 'fix'. Irrespective of the type of raid, robbery is one of the most serious risks to retailers. Even when violence is not used, the mere threat of it can have a dramatic affect upon staff. BRC statistics illustrate that, in terms of the total cost of crime, robbery (and till snatches) accounts for a very small percentage, at around 4 per cent. However, whereas the many thousands of other offences will result in little or no contact between staff and criminal, robbery brings staff into direct confrontation. Thus for the 14,000 robberies and 12,000 till snatches recorded by the BRC, there are at least as many, and probably more, victims. Indeed, Hibberd & Shapland (1993) maintain that robberies and till snatches represent a large problem for retailers and that repeat attacks often persuade shop owners to give up their businesses.

Fraud

The potential fraud problems faced by retailers can be broadly divided into three categories. The first category covers fraud by staff involving the embezzlement of funds or the taking of bribes. The second category involves fraud undertaken by customers, sometimes with the collusion of staff, using credit cards, debit cards and cheques. The final category of fraud encompasses the use of counterfeit bank notes and coins. However, as many frauds do not directly affect the retailer, rather financial institutions instead, they will not be considered in this chapter.

Violence against staff

The threat and use of violence is a serious problem for retailers and the example in November 1994 when John Penfold, a trainee manager in the Teddington branch of Woolworths, was stabbed to death during a robbery illustrates this (Union of Shop, Distributive and Allied Workers (USDAW) 1995). Hibberd and Shapland (1993) have also illustrated the problems faced by small shopkeepers, which range from physical assault by drunks and robbers to racial harassment in the form of verbal abuse and racist literature. Ekblom, Simon and Birdi (1988) have noted the particularly high incidence of threatening behaviour and verbal abuse encountered by Asian shopkeepers. For instance, one in seven of those surveyed had experienced assault. Poole (1991) has described the sexual harassment some staff face, from lewd comments to physical assaults. Beck, Gill and Willis (1994), in a survey of a national clothes retailer, found that 15 per cent of those surveyed had suffered a physical assault, 48 per cent had been sworn at and nearly 20 per cent had been threatened. In total, the 1998 BRC survey estimated that 11,000 staff were subjected to physical violence, 30,000 to threats of violence and 109,000 to verbal abuse. The survey also found the most common causes of physical violence against staff were staff attempting to prevent a theft or robbery and dealing with troublemakers, drunks and angry customers.

Blackmail

In recent years many retailers have become the target of extortionists. There have been attempts and examples of individuals putting glass in baby food or threatening to inject food with deadly viruses. Huge payments have been demanded in order to halt the extortion or disclose the location of contaminated products. In a typical example, a Mr Riolfo claimed he had AIDS and threatened to contaminate food products in a large supermarket chain unless he was paid £250,000. He was caught and jailed for eight years (*Daily Mirror*, 14 October 1995). More recently the 'Mardi Gras' bombers targeted Sainsbury's and a range of other organisations in an attempt to gain financial reward. The senior executives of some firms also face the risk of being kidnapped with ransoms demanded.

Terrorism

During the early 1990s PIRA changed its tactics and started bombing more public areas on the mainland. This affected retailers as some of the bombs were placed in shops or in shopping areas. Notable examples included the Baltic Exchange bombing in London in April 1992; fire bombs in city centre stores in Leeds and Milton Keynes in June 1992; the Manchester city centre bombings in April, May and December 1992; the Harrods bombing in January 1993; and the Bishopsgate bombing in April 1993 (Beck and Willis, 1994a). Since the end of the first ceasefire there were further bombings in the London Docklands and Manchester in 1996 and by dissident groups in Ireland. In addition to PIRA there have been increasingly militant actions undertaken by animal rights activists, including a number of attacks upon retailers linked to the 'meat' or 'fur trade'. The activities of these terrorist groups have combined to increase the fear of such an attack amongst shoppers. Research carried out by Beck and Willis (1994a) found that almost 62 per cent of those surveyed were worried about the prospect of an explosion and over 32 per cent had changed their shopping behaviour as a result. The 1998 crime survey recorded 3,000 outlets subject to terrorist incidents (defined as bomb threats and actual bombs), which amounted to costs of £10 million to retailers.

Risks to the retail community

There are some problems experienced by the retail community as a whole that may impinge indirectly or directly upon individual retailers. Some of these problems include public disorder, nuisance and the nature of the shopping environment. Terrorism could be equally applicable to this section as numerous PIRA attacks were not directed at one particular retailer, but rather a retail centre such as the 1996 bombing of the Manchester shopping area. However, as the vast majority of terrorist incidents were directed towards single retailers, terrorism was explored in the previous section.

Nuisance by groups who frequent shopping and town centres for almost every activity except shopping is very common. It ranges from vagrancy, fly-pitching, loitering and misbehaving to fighting (Phillips and Cochrane, 1988). Some of these activities are not necessarily criminal in nature, but they can lead to such behaviour. Nuisance affects the retail community collectively, for if a shopping or town centre gains a reputation for this sort of behaviour it will probably encourage shoppers to go elsewhere. Also, if a town or shopping centre gains a reputation for public order problems through drunks or football fans, similar consequences are likely to occur.

The environment of a shopping or town centre is also important in influencing the level of custom. Poole (1991) has identified a range of factors which increase the fear of crime. These include architectural fear generators such as: narrow access ways; restricted choice of routes; long escalators (where it is impossible to identify what is happening at the other end); large pillars (which could conceal an attacker); low ceilings; poor lighting in the shopping mall, car park or access routes; poor surveillance; and signs of decay such as excessive litter, vandalism and graffiti. Social fear generators include the presence of beggars, drunks, loiterers and the fear of being harassed by them. Most of these fear generators can also be applied to town centres, where there may also be other factors at work. For instance, Atkins, Husain and Storey (1991) and Ramsey and Newton (1991) have shown that poor street lighting can increase the fear of crime for vulnerable groups. If these factors are prevalent in a shopping or town centre they may make a shopper less likely to go there and this will ultimately have an effect upon the retailer's profits. A report on Nottingham town centre found the fear of crime led to a substantial proportion of women avoiding the city centre at night and a small proportion during daylight and at weekends (Nottingham Safer Cities Project, 1990).

The perpetrators of retail crime

This chapter has already touched upon the perpetrators of retail crime. The opportunist is a common offender. This may involve the overwhelming desire to shoplift where an individual suddenly finds him/herself in a shop with no staff, or the chance to embezzle when a lax procedure is identified by an accountant. Retail crime may also be undertaken by thrill seekers, particularly some young people who shoplift and cause criminal damage for pure excitement, often as part of their gang culture. Drug addicts may commit crimes such as robbery and shoplifting to pay for their habits. Due to their state of mind drug addicts and drunks may also assault staff. Criminal damage, assault and harassment against retailers of ethnic minorities sometimes have racist motives. More professional criminals may steal, rob and defraud to make a living. Organised crime involves robberies, large-scale frauds and counterfeiting operations. Finally, terrorists are engaged in a wide range of serious crimes, including bombings and arson, in order to raise funds.

The security strategies used to address retail crime

This chapter has demonstrated the numerous security risks retailers face and the huge costs involved. In response, retailers and other organisations have developed a number of strategies to combat crime at retail, shopping/town centre and national levels. Many of these strategies have also been pursued in partnership with other retailers, the police, local authorities and other bodies. The strategies pursued at a retail level by the larger retailers may also be part of a group security strategy.

The retail unit

Retailers have pursued a range of security strategies to address the problems of retail crime. The BRC estimated that retailers spent around £550 million on crime prevention during 1998. This accounted for the employment of security staff, the purchasing of specialised security equipment, the use of security procedures and a range of other measures, which will now be explored.

1. Security staff

Almost all the major high street chains have a security department or at least a security manager/ adviser. Often retailers employ security officers, store detectives and other specialist staff to undertake various security functions. They might also train other staff in security procedures. The presence of security officers patrolling stores has increased substantially in recent years because they are perceived to be more of a deterrent to shoplifters than anonymous store detectives. Indeed, research by Butler (1994) has shown that this is the case if security officers follow suspects around the store. They can also stop undesirables such as known shoplifters, beggars and drunks from entering the shop. They can assist store detectives with arrests as many of the latter are women who lack the strength to arrest determined shoplifters. Security officers can also reassure genuine shoppers and staff. Finally, they can also undertake various guarding duties in the store. Some retailers employ guards permanently to guard the stores at night because of the risk of burglary, criminal damage and arson etc.

Store detectives have been the traditional detectors and preventers of shoplifting and staff theft. They patrol a store in plain clothes looking for potential shoplifters and apprehending those caught in the act. They also undertake searches of staff to detect and prevent theft. Many store detectives are women and are either employed by a shop or group of shops in-house, or are contracted in from a security company (Murphy, 1986).

Some retailers also employ specialist security companies for a range of services. The most common is the use of a CIT company to transport cash and other valuables. For specialised fraud investigations and blackmail/kidnap situations, some retailers will also hire the services of a specialised private investigator. Senior executives who work abroad, and sometimes in the UK, may also require the services of a close protection company to prevent attacks or kidnapping, if they operate in a high risk area. All staff working for retailers have an important security role, whether it is spotting suspicious activities or discovering forged credit cards and notes. Therefore, the training of ordinary staff in basic security procedures is important in preventing and detecting retail crime.

2. Security products

Retailers use a wide variety of security products to prevent and detect crime. Surveillance security products are mainly used by retailers to prevent and detect shoplifting, although CCTV at least has wider uses. Mirrors have been used for many years by retailers to enable wider surveillance of 'blind spots' by shop staff. On the negative side, however, they can also be used by shoplifters to observe staff (Jones, 1997). In recent years CCTV has become the more ubiquitous and effective form of surveillance. It can be used to prevent and detect a wide range of retail crime, for instance: detecting theft by customers and staff, customers tampering with products, and staff involved in frauds. CCTV is also very useful for providing evidence from robberies, assaults upon staff and criminal damage. Perhaps most importantly, CCTV is an effective deterrent to some criminals. Butler (1994) found that shoplifters saw CCTV as the second highest risk to a successful theft. CCTV can also act to reassure customers in shops. Beck and Willis (1994b) discovered that 72.8 per cent of customers surveyed found it reassuring or very reassuring. The other main products used to aid surveillance are communication systems such as two-way radios, lighting systems, tannoy systems and bleepers. These systems are used by staff to draw attention to suspected criminal activities and causes for concern.

Alarms are also used by retailers to prevent a range of crimes. Intruder alarms are used to prevent and detect burglaries. Some retailers may also use loop alarms on some of the more valuable products, which are openly displayed. Loop alarms activate when a product is moved. Research by Beck and Willis (1994b) found that customers felt loop alarms were the biggest deterrent to shoplifters. Butler's (1994) research, based upon the testimony of shoplifters, found they considered loop alarms the fourth highest risk to them being detected. Some retailers in high-risk locations may also have panic alarms installed for staff to activate if they are attacked.

EAS, more colloquially known as tagging, is another form of in-store alarm system. Products have electronic tags attached and if they are not removed or deactivated at the point of sale an alarm is activated when they pass through a detection device at the gateway to a store. In a study of a national clothes retailers, Bamfield (1994) found that the introduction of EAS in the four stores assessed led to an average 28.3 per cent reduction in losses. However, Handford (1994) also notes some of the problems with EAS such as the ability of determined thieves to bypass the system and false alarms embarrassing genuine customers.

Retailers also use a range of physical security measures. Shops with very expensive and moveable stock, such as jewellers, make use of security cabinets to store products. Retailers may also make use of safes in which to store their products. Entrance to the shop is usually protected by locks. With the increasing use of the 'ram raid' and burglary by smashing the shop front window, retailers are also increasing their use of shutters, barriers and reinforced glass. Jacques (1994) found evidence that the use of shutters and barriers substantially reduces the incidence of 'ram raiding'. Banks and building societies are also increasingly making use of screens that rise when a cashier presses a button, thus separating staff from potential raiders in the event of a robbery.

3. Other security strategies

There are also a range of other security strategies used by retailers that do not fit easily into either of the other categories. These include the marking of property with ultraviolet ink to aid in the detection of thefts. A recent trend has been the installation of sprinklers that spray a unique dye or 'smart water' when an alarm is activated. This makes detection easier and deters some offenders from targeting stores that employ this security strategy. Some stores also make use of fake displays to show off their wares without the risk of the real goods being stolen. Signs are also used to warn that security procedures are in operation and that shoplifters will be prosecuted. Guard dogs are left to patrol some shops at night to deter burglary. Such is the frequency of assault, harassment and robberies in some small shops, that many have resorted to keeping weapons ready. Research by Hibberd and Shapland (1993) into small shops found almost half of those surveyed kept weapons available on the premises. With the increase in counterfeit notes many retailers are also using specialist equipment to detect forgeries.

Almost all retailers will have certain security procedures to minimise the risk of crime. A common strategy used by the larger retailers are exclusion orders. An individual caught shoplifting is banned from the store. If the individual persists in attempting to enter the store, retailers may then take court action to exclude them. Some of the larger retailers have their own internal systems of justice. Many are reluctant to prosecute certain groups of individuals committing shoplifting and use a system of warnings that ultimately lead to the bans described above. One of the more interesting findings is the reluctance of some retailers to prosecute for shoplifting. Nationally the 1995/96 RCS found that of the 1.6 million customers apprehended, only around 1 million were referred to the police. More recently some retailers have been experimenting with civil recovery where shoplifters are sued for compensation by retailers to recover their losses.

Many large retailers also have committees to discuss security strategies at a store, regional and group level. In some retailers there is a general store committee, which includes staff and management, with a wide range of responsibilities, one of which is security. Other stores have a dedicated security committee, consisting of management and staff representatives, which discusses how security can be enhanced.

The retail community

Retail crime can be addressed to an extent by individual retailers. More significant success, however, can only be achieved by partnerships within the industry combined with the police, local authorities and other relevant interests to develop strategies to combat crime. It is at this level that the differences in the success against crime can be distinguished between shopping centres and ordinary town centres, as the former are often designed with these strategies in mind. Almost all shopping centres and many town (or city) centres have managers who hold a wide range of roles, one of which is frequently the promotion and co-ordination of security strategies. CCTV systems, which cover town and shopping centres, have been one of the most common strategies and have expanded dramatically with government money funding these schemes. Often linked to CCTV systems, many centres have radio links where retailers are linked with radios so they can warn one another of suspected shoplifters. In some areas these have developed further, with the sharing of intelligence. Most shopping centres and many town centres also have their own security officers. Some centres also have drop-in centres for youths, which provide activities for them, and have been used in some centres to divert them from pursuing antisocial behaviour (Nottingham Safer Cities Project, 1990). Poole (1991) has also identified a number of factors that can reduce fear in shopping centres such as fresh flowers, designs that maximise light, and removing signs of decay. The management of joint strategies such as these requires specialist committees and in many centres these have evolved bringing together the different partners. Finally, the criminal justice system also plays a role through dedicated police units, probation officers and social services.

National strategies

There are also national strategies to combat retail crime. There are a number of national organisations that represent the interests of retailers, banks and building societies. Many of these have specialist committees/units dedicated to crime and security. These usually compile information on crime, analyse trends and disseminate information and intelligence to members. They also develop common security strategies and policies, while also attempting to influence the police and government. For example, the BRC has the Retail Crime Initiative (RCI). The RCI not only conducts the retail crime survey, but also seeks to influence the police, courts, Crown Prosecution Service and government with regard to retail crime. The BRC has also established a committee of all the major retail heads of security, which seeks to influence important bodies as well as sharing intelligence and best practice.

The government's Crime Reduction Unit (National Crime Prevention Agency as it was previously known) has a subgroup dedicated to retail crime. The Retail Action Group for Crime Prevention is tasked with finding new ways of developing and delivering crime prevention. The group's membership consists of various interests from the retail sector and other relevant interests. Its five key areas of work are: developing crime prevention training workshops for smaller retailers; the national promotion of crime prevention initiatives; tackling specific crimes; raising the awareness of commercial crime; and collecting more comprehensive data.

Conclusion

This chapter has elucidated the very wide range of risks facing retailers from the criminal fraternity and the high cost of crime to them. It has also shown that there are numerous security strategies that can be pursued to address these problems either nationally, by the retailer, or retailers together with other agencies. If the evidence of the RCS is accurate, retailers have enjoyed success over the last five years. In the 1992-93 survey the cost of crime to retailers was £2 billion and £2.15 billion in 1993-94. This figure was reduced to £1.5 billion in 1994-95, £1.42 billion in 1995-96, £1.38 billion in 1997 with a slight increase to £1.39 billion in 1998. The retail sector has been an example of how effective security strategies can be employed to tackle crime. However, while the risks in some areas may diminish, new challenges will emerge that must be addressed as, inevitably, criminals will find new ways to combat the successes of retailers, thus fuelling the security 'arms race'.

Chapter 14

Aviation Security

The aviation industry is one of the most security conscious sectors of commerce with some of the most serious security risks. As a consequence the aviation industry is subject to stringent regulations covering security. The range of security risks and the strategies that have evolved to meet them make aviation security one of the most idiosyncratic and interesting sectors to study. Security concerns range from the hijacking of aircraft and mortar bombing of airports, down to the more familiar risks of theft, fraud and robbery. In light of these unique and extremely dangerous risks, governments and international organisations have intervened and stimulated security strategies and procedures that are at the very cutting edge. This chapter will examine the range of security risks specific to the aviation industry. It will then analyse the rationale for and extent of international and government intervention. The security strategies that have emerged to combat these security risks will then be considered, before potential future trends are assessed. The aviation industry, for the purposes of this chapter, will consist of airports, and passenger and cargo carrying airlines.

The range of security risks to the aviation industry

Attack upon aircraft/airport

There are a number of scenarios by which an attack could take place against an aircraft or airport. An aircraft could be attacked in the air by another aircraft or from a missile fired from the ground. A plane could break up as a result of an exploding bomb that was carried in luggage or cargo (this type of risk will be examined separately later in this section). An aircraft could also be attacked on the ground by mortars, missiles or even small arms fire, as could an airport. Furthermore, an attack does not necessarily have to involve a weapon; it could be simply an intruder sabotaging an aircraft on the ground. These incidents, although gaining a very high profile, are very rare. In most years since 1947 there have been no recorded incidents of this nature and between 1947 and 1993 there were only 62 incidents, with the most occurring in any one year 4 (CAA, 1976 and 1994). Unfortunately there are no readily accessible statistics relating to attacks on grounded aircraft or attacks directed against airports, although these incidents are rare. Perhaps one of the most prominent British examples was a spate of mortar attacks by PIRA on Heathrow airport during March 1994.

Attack upon aircrew/passengers

Random attacks upon passengers at an airport are extremely rare. In probably the most infamous examples, the Rome and Vienna airports suffered simultaneous terrorist attacks in December 1985. Terrorists at the airports opened fire randomly with AK-47 rifles and hand grenades on passengers waiting to check in. The result was 15 dead and 74 wounded at Rome, and 3 dead and 47 wounded at Vienna. There have also been a number of attacks upon aircrew and passengers in flight, more often than not during a hijack, which will be discussed separately later. However, some attacks upon aircrew in flight could not be unambiguously described as the result of a hijacking. For example, in December 1987 a sacked employee from USAir used his airline identification card to bypass security and board a flight. Once in the air he shot a customer services manager from USAir and then the crew, causing the plane to crash killing 43 people (Moore, 1991) (This has not been classed as a hijacking because the motive of the individual appears to have been to attack his former colleagues, not to seize the aircraft for ransom or publicity). The CAA World Airline Accident Summary keeps statistics of the number of crew shot while working. One crew member was shot in each of the following years: 1948, 1952, 1954, 1966, 1971, 1977, 1987, 1988, 1989 and two crew members were shot in 1970 (CAA 1976 & 1993).

The increasing length of some flights combined with the free and plentiful supply of alcohol has led to an increasing number of in-flight brawls and related incidents (House of Commons Official Reports, Standing Committee B, 6 June 1996). During the passage of the 1996 Civil Aviation Amendment Act, which plugged a gap in the law to give jurisdiction to British courts for offences taking place on international flights, in international airspace, heading to the UK, a number of incidents were mentioned. These incidents included sexual assaults, brawls, drunkenness and couples engaged in sexual acts in public. One of the best illustrations of the problem was the case of one family from Kilburn in London on a flight to the USA. The family started a drunken brawl and were only subdued by members of the US wrestling team, who were fortunately on the flight (House of Commons Official Reports, Standing Committee B, 6 June 1996).

Bombs

Probably the most cowardly of all potential attacks upon an aircraft or airline is the bomb. A bomb or bomb hoax at an airport or on airline can cause damage ranging from panic, to large scale loss of life. In June 1980 a bomb exploded in a locker in Orly airport in France injuring six. In 1985 a bomb exploded in a suitcase at the Leonardo da Vinci Airport in Rome wounding 15 employees. Worst of all, however, is the cowardly act of placing a bomb on an aircraft. With just a small explosive detonated at the right moment it is capable of causing massive loss of life. This was tragically demonstrated on the Air India flight from Montreal to London in 1985 (329 casualties) and the Pan Am Flight 103 in 1989 (269 casualties, including 11 on the ground). Terrorists have increasingly resorted to this instrument of terror and since 1969 there have been 70 known attempts to plant bombs on board airlines, which have caused at least 15 crashes and 1,732 casualties (Jenkins, 1999)!

The explosives are deposited on an aircraft in a number of ways. The most common means is for an explosive to be placed in luggage and then checked in by an unsuspecting passenger or suicide bomber. Another method is for a terrorist to check in the luggage, but not take the flight. A more ingenious method, which recently led to the death of a passenger on a flight from the Philippines, was allegedly used by Ramzi Youssef when he took a bomb on board in pieces and built it in the toilets. He then placed it under a seat and left the aircraft before the bomb exploded on the next leg of the journey. A bomb could also be placed on a plane by a corrupt member of staff.

Drug abuse by aircrew and air traffic controllers

The dangers of aircrew, air traffic controllers or other personnel with safety related roles working while under the influence of alcohol or drugs are clear to all. An aircraft being flown by a pilot under the influence of alcohol could increase the chances of an accident. There are strict rules in most countries with regard to the consumption of alcohol and drugs. Some countries, including the USA, have also moved towards random drug testing.

The extent of abuse of this kind is difficult, if not impossible, to gauge because many abusers go undetected. Furthermore, many cases are not publicised, for obvious reasons.

In one sensational case in March 1990, however, three Northwest Airline pilots operated a flight while under the influence of alcohol. The captain was reported to have drunk 15 to 20 rum and cokes the night before his flight was due to depart. He was so drunk that he had to return to the bar to ask his way back to his hotel, which was only three blocks away. Blood tests showed the pilot had the equivalent of eight drinks in his system when the plane departed the following day (Moore, 1991)! A television programme on Aeroflot, the old Soviet national airline, also revealed tales of pilots drinking vodka before take off (True Stories, broadcast on Channel 4, 13 July 1995).

Counterfeit air parts

In recent years there has been an increase in the number of fake and bogus air parts for aircraft. The use of such parts has obvious dangers for an aircraft where specially made, intricate and strengthened parts are the norm. The use of parts that do not meet safety standards puts planes at risk. For example, Flight PAR 394 from Oslo to Hamburg took off in September 1989 and broke up half an hour after take off, killing 50 passengers. It was subsequently found that the tail had broken off because the bolts that had been fitted were bogus and not manufactured to the special standards required. In another example, a spare parts dealer was raided in the US and numerous matresses were found. The bed-springs in the matresses were being taken out and used for a part in the engine starter assembly! Expert testing revealed these starters could have resulted in the engine exploding (Panorama, broadcast BBC1, 12 June, 1995)!

Hijackings

The very first recorded act of hijacking was in 1930 when Peruvian revolutionaries seized an F7 Fokker aircraft. This marked the beginning of an era for this new crime, but it was not until the post-war period, with the huge growth in air passenger transport, that hijacking came to international prominence. There is much debate on how to define hijacking, although the essence of it is the unlawful seizure of an aircraft. The risk of hijacking is almost totally a threat to passenger airlines, although it is also possible to hijack a plane carrying cargo. Phillips (1973) divides the post-war period into two with regard to hijacking. The first half, 1947-1958, he calls the propeller years, when hijackers were predominantly Eastern Europeans seeking to escape Communism. All but three hijackings in this period took place in Eastern Europe. The second half, 1958-72, he calls the jet age, when hijacking became a worldwide problem and the motives behind it became more diverse.

More recent research has divided the number of hijackings into five decades. Between 1947-56 there were 17, 1957-1966 there were 37, 1967-76 the number peaked at 385, declining to 300 during 1977-86, and then 212 between 1987-96 (Merari, 1999). The peak between 1967-76 can be largely accounted for by exiled Cubans attempting to return from the US (Dorey, 1983). For instance, 19 of the 22 hijackings in the USA in 1968 were directed to fly to Cuba. By March in the following year, there had already been 19 'diversions' to Cuba! Such was the extent of the problem that airport authorities, at Christmas, were reminded to look for suspicious Cubans! However, the beginning of modern terrorist hijacks can be charted to the events of September 1970 when terrorists hijacked a TWA 707 from Frankfurt to New York. On the same day Palestinian terrorists hijacked a Swissair DC-8 and a Pan Am 747.

A few days later a BOAVC VC-10 was also hijacked completing the quartet. The Pan Am aeroplane was forced eventually to Cairo where it was blown up, while the other three were forced to Jordan where they were also blown up (Moore, 1991). Since 1970, the hijack of an aircraft by a terrorist group has become a familiar occurrence. Such attacks have stimulated national, as well as international, action to try and combat them, and these initiatives will be discussed later in this chapter.

Illegal aliens

Illegal aliens are a risk to both cargo and passenger airlines. An individual may try to escape from a country by hiding in the cargo hold or simply by travelling as a passenger on a forged visa and/or passport. There is little risk to the security of an airline if the individual is merely trying to escape to another country, (although there is a risk to the safety of the individual if they are hiding in the hold). Under the 1987 *Immigration (Carriers Liability) Act,* however, an airline (or ferry operator) may face a fine of £2,000 if a passenger has either invalid or forged travel documents when he or she arrives at the end of the journey (NAO, 1995). In 1992 it was estimated that fines of over US$50 million were imposed upon airlines for alleged visa and document irregularities. In addition, the costs of legal fees, repatriation of passengers, accommodation and subsistence, management time and poor publicity may also be added to this figuure (Aviation Defence International Promotional Literature). It is therefore essential for companies to pursue measures to reduce this risk.

Intruders

Intruders are mainly a threat to an airport but it is conceivable that they could also pose a risk to an aircraft. They can be divided into accidental and criminal intrusions. The risk of an individual either walking or driving on to a runway by accident is clear. If combined with an aircraft landing it could cause a severe accident. The risks posed by accidental intruders are very slight given modern security measures protecting an airport. Nevertheless, it has been known for children to cycle on to runways in less secure airports and families to have picnics in the grounds (Dorey, 1983). There is also the risk of wild animals getting into an airport.

The more likely intrusions are usually criminally motivated. The large amounts of expensive equipment at airports may tempt criminals to penetrate the security defences, with the aim of stealing from hangars and aircraft. Intruders may also be individuals with more sinister motives in mind. In January 1975, for instance, Paul Landers climbed the perimeter fence at Pensacola Airport in Florida and attempted to hijack a National Airlines B 727. Unfortunately for Landers the only staff on board the plane were cleaners (Dorey, 1983). Recently in Great Britain more common threats have come from groups protesting at the live export of calves. In February 1995 planes were delayed when protesters broke through a fence at Coventry airport to campaign against the live export of calves (*The Guardian*, 4 February 1995).

Ordinary risks

Airports and airlines, like any organisation, face a range of criminal risks. Arson, burglary, criminal damage, fraud, kidnapping, robbery, theft — particularly of luggage (by customers and staff), and violence against staff, are all threats that any airline or airport will face, in addition to the more aviation specific risks. Some of these crimes will occur in a different context to a typical organisation. For example, stealing an aircraft poses a very different set of problems to stealing a car for both the security manager and the criminal. Conversely, other crimes, such as fraud, pose the same problems for an airport/airline as for any other large organisation. As many of the ordinary risks are similar to those faced by other sectors of commerce, they have not been reviewed in detail here, but similar risks and strategies apply.

The origin of risks

The risk to aviation security comes from a wide range of individuals and organisations as the preceding section has shown. By far the greatest risk is from terrorists who have been responsible for the vast majority of hijacks, bombings and attacks. Organised criminals and petty criminals have also been very active in affecting the aviation industry through their involvement in thefts, robberies, smuggling contraband and even hijacking a plane for the purpose of robbery. Some criminal acts in aviation have been undertaken by the insane or disturbed, while others, particularly hijacks, have been perpetrated by those desperate to flee to another country. However, it is terrorists who have posed the most dangerous threat and it is in order to deter and prevent their activities that the international community, national governments and the industry have developed a number of strategies.

Strategies to improve aviation security

A wide number of strategies have been pursued to enhance aviation security. The nature of aviation has also meant that there has been a large degree of international involvement in developing security strategies through international governmental organisations and associations. National level responses have also emerged through national governments, regulatory bodies and trade associations. These have all combined to help co-ordinate and influence a wide range of security strategies used by airlines, airports and security companies.

The international response

The increasing risks of terrorists and others disrupting air travel over the last 30 years have led to a number of initiatives to combat the problem, many originating at an international level (see Wallis, 1999 for an extensive account of international measures). Due to the very nature of air travel, countries cannot tackle this problem in isolation. If minimum security standards are laid down in state x, but not in state y, then for flights between x and y the minimum standards set by x are useless. Therefore, in order to address the problem it has been essential that aviation terrorism is tackled at an international level. Some of the most important organisations that have been involved in this process include the International Civil Aviation Organisation (ICAO), which was founded under the Chicago Convention of 1944 and is an agency of the United Nations; the International Air Transport Association (IATA), which is an international trade association representing airlines; and the European Civil Aviation Conference (ECAC), which is a regional grouping of ICAO. The Federal Aviation Administration (FAA) of the USA is also a very influential body at an international level because of the world dominance of US aviation.

The international legal framework regulating aviation security has been established by a number of conventions and agreements. The Chicago Convention on International Civil Aviation of 1944 is important insofar as it established ICAO. However, the first convention that really tackled the issues of aviation crime was the 1963 Tokyo Convention on Offences and Certain Other Acts Committed on Board Aircraft. This was the grandfather law, establishing rules on jurisdiction of states with regard to offences committed in international territory, and the powers of a commander of an aircraft. The legislation also established a framework to deal with unlawful acts on board aircraft and the penalties for offenders. Interestingly, there are just two paragraphs dealing with the Unlawful Seizure of Aircraft (Chapter IV, Article 11).

The increase in hijacking towards the end of the 1960s led to the realisation that the international framework was inadequate. The response of the international community was the 1970 Hague Convention for the Suppression of Unlawful Seizure of Aircraft. It specified what states should do when there is an unlawful seizure, called for heavy penalties for those convicted of hijacking and provided for the extradition of hijackers.

As the Hague Convention was being drafted, the realisation dawned that it did not deal with acts of sabotage. So, rather than add to the Hague Convention another was drafted. The result was the 1971 Montreal Convention for the Suppression of Unlawful Acts Against the Safety of Civil Aviation. This Convention established the framework for acts of sabotage against aircraft. It was subsequently widened to include airports, after the Rome and Vienna airport attacks, with the Protocol for the Suppression of Unlawful Acts of Violence at Airports Serving International Civil Aviation, Supplementary to the Montreal Convention 1971. The Montreal Convention also included a very important clause in Article 10. This clause allowed ICAO to develop Annex 17, which explains the standards of security that are required to 'prevent offences'. In order to achieve this, Annex 17 has standards, that are 'mandatory' and practices, that are optional. These standards and practices are the result of negotiation between the experts and the politicians. However, the mandatory rules can be opted out of (Wallis, 1993a).

The international framework to tackle the problems of terrorism seems adequate on paper, but it suffers from the problem all international agreements face: enforcement. By the mid-1970s it was clear to many countries that some signatories were not operating within the spirit of the various Conventions and some were even working against them. Some states were not implementing the measures either for financial reasons or simply because they had no intention of complying. Other states were failing to extradite terrorists or imposing slight penalties upon them. Certain regimes were even sponsoring terrorism. At the Group of Seven Bonn summit in 1978 the world's leading industrial countries made a renewed effort to address these problems. The 'Bonn Declaration', which was issued at the summit, stated that all flights to and from those countries that failed to honour the spirit of the Conventions would be prohibited. The declaration, however, has rarely been used, with threats of its imposition against South Africa, and its actual use against Afghanistan, the only examples.

This framework has thus established penalties for criminal activities in aviation to deter potential offenders, and minimum security procedures in order to prevent such actions. The international framework sets the broad principles, but the fine details of the standards required under the Conventions are established through regulations issued under national governmental legislation. The following figure illustrates how this process works in the UK.

Figure 5. The implementation process for international agreements: a case study of the UK

Note: ECAC can and does operate independently of ICAO in establishing security standards.

The figure illustrates the process from international agreement through to company security procedure. An international standard is agreed, either at ICAO (Annex 17) or ECAC (Document 30) level. There may also be input and recommendations from other international organisations, such as IATA, at this stage. This leads to governments introducing national legislation (although the legislation may already be in force, which allows the regulatory body to reissue regulations under existing legislation). In the case of the UK rules are implemented through the National Aviation Security Programme (NASP), which is organised by TRANSEC (the British regulatory body responsible). This is a document that lists which security strategies are required from organisations, most of which are legally binding. These standards are then issued to the company and they must maintain them. If the firm in question uses a contract security company, then these standards are set through a contract with them. The standards that evolve at each level may expand. A company may wish to have higher standards than the imposed minimum, and a national government may also wish to impose higher standards than those specified by the international body.

These standards are enforced through a national inspectorate, which in the UK is TRANSEC. The inspectorate conducts audits of a company's entire security procedures and a detailed examination of a single aspect. It also tests the simulated attempt to defeat a company's security procedures. Those failing to meet these standards face disciplinary action. In addition, companies have their own auditing procedures and may, if using contract security, test them as well. In the UK, TRANSEC is also responsible for regulating security for the railways, the Channel Tunnel and the ports. The arrangements differ in other countries. In the USA, for example, the FAA (which is part of the Department of Transportation) is responsible for aviation security. If compared to the UK, the FAA has the role of both the CAA and TRANSEC (excluding non-aviation security).

Security strategies developed and mandated to prevent aviation terrorism

TRANSEC mandates a framework of security strategies to combat security risks to aviation. Combined with the many compulsory security procedures, there are also a variety of other strategies that have emerged. Some of the more important strategies and procedures that have been established will now be examined. Central to many of these strategies are the security officers who implement them.

1. Screening and searching of passengers, luggage and cargo

To prevent individuals smuggling guns on board an aircraft or placing bombs in luggage, screening and searching procedures are used. Usually all passengers pass through a metal detector and they may also be screened by a security officer with a magnetic wand if there are any doubts raised by the initial screening. Some passengers may even be subject to a hand search. Hand baggage and luggage may also be subject to various searches and screening procedures often using specialist equipment (Baldeschwieler, 1993).

More extensive searches may also be conducted by security officers at airports. These may include the searching of an aircraft before passengers board, the searching of vehicles entering the airport and the searching of the grounds of the airport. Indeed, the Secretary of State has the power to enforce searches at aerodromes (Section 13 of the 1982 *Aviation Security Act* as amended). Some airlines/countries also insist on passenger profiling, where an individual is asked a series of questions when checking in. Depending upon the answers, the security operative specifies certain action with regard to the screening and searching, that should take place for that person's luggage.

2. Secure areas

The manager of an aerodrome can apply to the Secretary of State for part or all of an airport to be a restricted zone (Section 12 of the 1982 *Aviation Security Act* as amended). Almost all airports are divided into airside and landside. The former can be further divided into areas such as the sterile concourse, where passengers wait to board their flight once they have passed through the screening procedures, and the ramp, where the plane waits for passengers to board. To transfer from landside to airside it is necessary to either have a valid pass, or to go through the security screening procedures, and sometimes both. The securing of these objectives is achieved through a number of security procedures. These include access control systems, CCTV, fences, barriers, security officers and intruder alarms to name the most frequent.

3. Passenger/baggage reconciliation

In recent years terrorists have increasingly switched tactics from the hijacking to the bombing of aircraft. This has posed a problem to security experts as there is no technology currently available that would detect 100 per cent of explosives hidden in luggage (Wallis, 1993b and Oxley, 1993). As most terrorists would not want to be on the plane when it explodes and Air India in 1985 and Pan Am 103 in 1988 were the result of unreconciled luggage on board, one of the strategies pursued has been a 100 per cent reconciliation of baggage. Thus, if a passenger checks in and his/her luggage is put on the aircraft but then fails to board the flight, the luggage is not carried. This, however, does not deter the fanatic or the suicidal.

4. Cargo security procedures

The transport of cargo presents a different set of challenges to the security manager. The threat of attack may be lower than for a passenger carrying aircraft but there is still a risk. A bomb could be placed in cargo and detonated when the plane is over a city and the carnage could be just as terrible (Litherland, 1994). To address these potential problems regulations concerning the transportation of cargo have been introduced. Under the British rules *(The Aviation Security [Air Cargo Agents] Regulations* 1993) if an organisation exporting cargo applies for and meets the standards to become a Listed Air Cargo Agent, then the cargo does not have to go through additional security procedures at the airport. In contrast, cargo from a non-listed agent has to go through these additional procedures. To become listed the organisation must meet a range of security standards on screening, vetting and training of staff, and security of buildings and transportation (Department of Transport, 1993). The minimum standards of security have therefore been extended beyond the airport to premises and employees of those organisations that use aircraft to transport their goods.

5. Raising the security standards of staff

It is essential that if the security strategies mentioned are to succeed the personnel operating them are of good character and are not a security risk. It is also vital that the staff operating these strategies understand why they are required and have the skills and knowledge to make them work. Minimum standards of vetting and training have been established in the UK to achieve this objective for security staff, as well as aircrew, and senior airport/airline managers, amongst others. These standards are set and enforced by TRANSEC.

6. Airport security committees

The NASP in the UK also requires that aerodromes establish an Airport Security Committee (ASC). The ASC has the role of advising on security matters, maintaining a list of vulnerable points, recommending appropriate security measures and reviewing contingency plans. The NASP also sets down requirements concerning the airport security manager's responsibilities. These include ensuring the protection of the airport's and the company's assets, providing a liaison point with the Department of Transport and even keeping abreast of technology with regard to security. The airport security manager is also expected to establish an Airport Security Programme, which sets out NASP's policies for the airport.

7. Other

Given space constraints, it is not possible to list every security strategy that has been used and every mandatory security procedure required, although the most important have been discussed. However, it might be useful at this point to briefly mention some of the other bodies involved in aviation security beyond private security. As many of the offences against aviation are perpetrated by terrorists, the forces that are involved in the fight against them are also relevant. The police Anti-Terrorist Squad, the Security Service and the armed forces all have a role in combating aviation offences in the UK. The police maintain law and order at airports. They also conduct armed patrols in a number of UK airports. The Anti-Terrorist Squad and Security Service compile intelligence, which influences the level of airport security. The armed forces have been used to prevent mortar attacks on UK airports, and the Special Air Service might be used to storm a hijacked aircraft if necessary. The organisations involved in other countries may be different but the same functions and activities are also carried out.

The effectiveness of security strategies

The problems in enforcing the international Conventions have already been mentioned and this is a major barrier to the success of aviation security. A number of other problems have emerged, however, in maximising the effectiveness of aviation security. Among the strongest indicators of the effectiveness of security standards are the tests undertaken by regulatory agencies. When the FAA introduced its monitoring programme in 1987 to screen passengers for weapons, it found 20 per cent of those attempting to take weapons through security avoided detection. Indeed, during the first two sweeps the FAA imposed $2 million of fines on security operators (Wallis, 1993b). Another report was equally critical of US airport security after finding that inspectors could gain access to secure areas as unauthorised persons on 75 per cent of all occasions (US Department of Transport, 1993). Of course these are only specific examples, but they do not inspire confidence in aviation security measures in the USA.

In the past, security strategies used have failed to keep up to date with the threats posed. Wilkinson (1993) cites the example of the late 1980s when the threat of hijack was receding, but the threat posed by bombs was increasing. The international community only responded to this new threat after a number of tragic incidents. The motivation and pay of security staff are also significant issues, as generally in the UK private security officers employed at airports are relatively poorly paid. No matter how good the technology, ultimately it depends upon the skills of the security staff. Operating the latest screening technology will have little impact if the security personnel watching it have little motivation and just rush the luggage through so they can have a longer break.

There is also a public perception among aircraft passengers that the security standards are inadequate. A survey by Manto (1993) found that 59 per cent of those surveyed disagreed or strongly disagreed with the statement that 'Airline Security in the US is Adequate'. He also found that nearly half would be willing to endure longer processing times, although only just over a third would be willing to pay more. If these results are representative, then it seems aviation security has a credibility problem and large sections of the travelling public would be willing to endure more intrusive security measures to achieve greater safety.

Despite the evidence to illustrate the inadequacies of aviation security there have been some positive results. In 1974, the year after legislation was introduced for mandatory US security measures, the number of attempted hijackings was reduced from 29 to 3. In addition, 3,500lbs of high explosive, 2,000 guns and 23,000 knives were detected (Wilkinson, 1993). The example of El Al illustrates that if the public is willing to endure increased security procedures and there is a commitment to tough and rigorous security, then there can be an effective system to deter and prevent attacks. The overall assessment, however, is not as good, as Merari (1999: 23) has argued:

> Despite the long accumulated experience with attacks on commercial aviation, and not withstanding the immense investment in security measures and procedures, the effectiveness of aviation security measures has not improved during the past three decades.

Conclusion

The one certain factor with regard to the future of aviation security is that terrorism will not disappear (Wilkinson, 1996 and 1999). Some groups may disband, be defeated or give up arms, but there will always be another group somewhere in the world that is willing to use terror. As long as this continues aviation will be a potential target for the terrorists because of the easy publicity such attacks can bring. The security strategies that emerge may make an attack more difficult or deter it, but this will merely encourage new technological advances or deflect terrorists elsewhere. It is therefore the challenge of security experts to keep one step ahead.

Security experts have predicted a number of threats that the aviation industry may face in the near future. One of the more common perceived future threats are the use by terrorists of missiles (McGuire, 1989; and Shaffer, 1999). The use of chemical and biological agents by terrorists to attack an aircraft or aircraft terminal are other potential risks for the future (Wilkinson, 1999), which has already occurred in other transport locations. For example, in 1995 members of the Aum Shinrikyo (Supreme Truth) used sarin gas on the Tokyo Metro killing 12 and injuring hundreds. The development of plastic weapons that are not detected by screening procedures could also lead to the resurgence of hijacking as terrorists would be able to smuggle weapons on board. Women may also be increasingly used by terrorist groups as they might be viewed as more capable of circumventing security procedures than men (McGuire, 1989). There have also been attempts to bring down an airliner by using false air traffic control signals (*Sunday Mirror*, 24 September 1995). It is not unimaginable that a terrorist group or mentally unstable individual may use this as a means of attack in the future. The future of aviation security holds many new and existing threats and the international community must be ready to meet these challenges.

Chapter 15

Entertainment Security

Entertainment is a nebulous word, the essence of which is amusement, hospitality and public performance. A fairly comprehensive list might include sports such as football, rugby, cricket, tennis, golf, horse/greyhound racing, boxing, athletics, swimming, snooker and motor racing. It might also include arts events such as plays, films, ballets, operas, and classical and pop concerts. Places of recreation such as pubs, restaurants, nightclubs, comedy clubs, bookmakers, bingo halls and casinos might also be included. An additional category might include special events and exhibitions, agricultural shows, antique fairs, motor shows, carnivals, firework displays and beer festivals. These activities would all be regarded by most people in varying degrees as forms of entertainment. Given the size of this sector the discussion will be restricted to the more popular forms of entertainment. The analysis will consider the security risks in the entertainment industry and where the risks come from. We will then discuss the general strategies that have developed to address some of these risks.

The security risks to entertainment

Security risks in the entertainment industry are extremely diverse. Certain risks are restricted to specific types of entertainment and some may even be exceptional to one team or group. For example, some football teams attract greater levels of hooliganism than others do. Certain pop groups also attract a following that causes more complications than others do. Some of the more common security risks to entertainment will now be explored.

Attacks on entertainers

In most forms of amusement there are entertainers, whether it is pop stars at a concert, footballers on a pitch or actors on a stage. Many entertainers are very famous, often extremely rich and frequently inspire great excitement in those who that watch them. This leads to a number of security risks for those who have responsibility for their protection. Clearly the worst risk would be an attempt to murder or seriously injure an entertainer. This is rare but such attempts have occurred. For example, a fan of the pop singer Bjork recently sent a letter bomb to her, which was luckily intercepted and defused. Sometimes stars are not as fortunate — John Lennon was murdered by an obsessed fan, and an attempt on the life of other ex-Beatle, George Harrison, also recently occurred. Jill Dando, a television presenter, was also murdered and her alleged killer is currently awaiting trial.

The more common attacks on entertainers are not as serious, but they can cause problems. There have been a number of attacks on players by 'fans' at professional football matches. Players have been hit by missiles thrown from the crowd and in some cases assaulted. During a football match between Millwall and Sheffield Wednesday a fan invaded the pitch and threatened the Sheffield Wednesday goalkeeper, Kevin Pressman (*The Guardian*, 26 October 1995). There have also been a number of attacks on match officials at sporting events. At a Portsmouth versus Sheffield United match in 1998 an assistant referee was knocked unconscious by an irate Sheffield fan. It is not only football where sportsmen risk being attacked by those that are supposed to be watching them. The fight between Riddick Bowe and Andrew Golata in Madison Square Gardens provides a notable example. After Golata had been disqualified by the referee for punches below the belt, he was attacked by one of Bowe's camp. This attack triggered off a mass brawl between rival supporters (*The Guardian*, 13 July 1996).

Attacks by entertainers

Often entertainers are the risk themselves. At the 1996 Snooker World Championships, Ronnie O'Sullivan, one of the leading players, was alleged to have attacked a tournament press officer. The history of pop music is also full of bands brawling amongst each other. The Gallagher brothers of Oasis have been the most recent example and the band's 1996 US Autumn tour was cancelled mid-term because of internal disagreements. Indeed, in an interview Noel Gallagher claimed the band's security was employed primarily to keep him and his 'kid brother' apart (*Sunday Times Style*, 15 September 1996). Many pop groups have also been notorious for causing serious damage to their surroundings, particularly hotel rooms. The Who were famed for their destructive capabilities, whether it was their equipment on stage or the furnishings in their hotel. During one typical spree in a New Zealand hotel in 1968, most of the furniture went out the window. In another example at the 1968 Tracks Record Christmas party, which was attended by The Who and The Small Faces, a food fight started resulting in a 'maelstrom of food' (Welch, 1995).

It is not only property and fellow entertainers that are at risk, spectators can be as well. Eric Cantona, the former Manchester United footballer, launched one of the most famous attacks on a spectator after being sent off against Crystal Palace. Cantona lunged at a Crystal Palace fan who had been abusing him. In a non-sporting example, Axl Rose, the lead singer of Guns N' Roses, was arrested in 1992 after leaping into the crowd and attacking a member of the audience. During the late 1970s some punk bands were also notorious for spitting at their audiences.

Disruption of entertainment

In most popular sports and some other forms of entertainment there is a risk of disruption by the crowd. There are a number of ways in which this could occur, but in this section we shall concentrate on streakers and protesters. Streaking appears to be a peculiarly British phenomenon that has occurred in virtually every popular sport. During the 1996 men's tennis final at Wimbledon the first ever streaker managed to find her way on to the centre court. In the same year, streaking finally arrived at polo when two individuals bared all at Cartier International Day at the Guards Polo Club at Windsor (*Daily Telegraph*, 29 July 1996). Streaking is probably most frequent at cricket matches, but it also regularly occurs during football and rugby matches, and even horse racing meetings. It is not dangerous, but streaking does disrupt the sporting event concerned. In numerous cricket matches streaking has often resulted in the batsman getting out to the next ball!

Another source of disruption for sporting and other events are demonstrations by protesters and crowds moved to spontaneous action. Demonstrations have affected a number of sports. Horse racing has a long history of attracting protest, most notably the suffragettes at the beginning of the twentieth century. Animal rights protesters caused disruption to the 1993 Grand National, which contributed to its failure to take place. During the early 1970s there was severe disruption of a number of sporting events by protesters campaigning against tours by South African teams to the UK. Major sporting events have also been targeted by terrorists. The 1997 Grand National was postponed because of a PIRA bomb warning.

At football matches there are often spontaneous as well as planned protests by fans. Often they are protesting over the selling of a star player, calling for the sacking of a manager or for some other grievance. At the end of the 1994-95 season fans at Brighton and Hove Albion invaded the pitch causing the match against York to be abandoned. The fans were protesting at the decision of the board to sell the Goldstone Ground, despite the lack of a new home stadium. Pitch invasions also occur in cricket matches. One of the most notable examples was the 1996 under-15 match between Pakistan and India at Lords. With six runs required for victory by India, 200 Pakistan fans invaded the pitch causing the Indian players to seek refuge in the pavilion and the match to be abandoned. Even motor racing attracts protesters. At the 1996 Australian Grand Prix in Adelaide, £1 million was added to the security bill, after threats by protesters to disrupt the event in protest at the race track being built on parkland. Disruption is not only confined to sporting events. At the 1996 Brit Awards while Michael Jackson was performing, Jarvis Cocker from Pulp ran on to the stage while Michael Jackson was performing in order to disrupt his performance.

Disorder

Disorder is frequent at a number of places of entertainment. On most Friday and Saturday nights both inside and outside pubs and nightclubs, there are fights involving drunken men and women. Probably the most salient examples of hooliganism, however, occur at sporting events, particularly football matches and to a lesser extent boxing and cricket matches. Football hooliganism is not a new phenomenon and it is also a problem that has been the subject of much research (Williams, 1985). Great strides have been made in tackling the problem during recent years, but outbreaks of hooliganism still occur. After the defeat of England by Germany in a penalty shoot-out in the 1996 European football championships, the frustration of England fans spilled into hooliganism throughout the country. The problem of football hooliganism has been acute in the past, not only leading to the disruption of matches but even to the deaths of fans. These problems, however, pale in comparison to countries such as Lebanon, where football matches between rival militias are policed by the army with fixed bayonets, and troublesome fixtures are held in empty stadiums!

Several recent boxing matches have resulted in widespread disorder. This can be partly attributed to the spectacle of two men fighting in front of a crowd of often drunken working-class males coupled with a partisan following for each boxer. The September 1994 fight between Birmingham boxer Robert McCracken and his Manchester rival Steve Foster, at the Birmingham NEC, resulted in clashes between rival fans. At another fight between Jim Murray and Drew Docherty, rival fans fought and rampaged through the venue while the former lay dying. Such disorder at boxing matches is not unique to the UK as discussed earlier when discribing the mass brawl at the fight between Andrew Golata and Riddick Bowe in New York.

Even cricket in recent years has begun to suffer from disorder or at least drunkenness. At the 1996 Test match between England and Pakistan at Headingley there were a number of clashes between rival fans. Crowd disorder has even emerged on the Grand Prix circuit. During the 1996 British Grand Prix a camp of German fans was attacked by a group of around 30 British fans.

Racism

Some form of racism probably exists in most forms of entertainment in varying degrees. It is in sport, and in particular football, where racism manifests itself most prominently. It takes the form of chanting racist slogans, making derogatory noises, such as those of a monkey, and throwing racist symbols, such as bananas, at players. Research of fanzine editors of football clubs has shown that 79 per cent felt that racism was still in the game, 51 per cent had witnessed mass racist chanting and 21 per cent had witnessed extreme right-wing activity (Garland and Rowe, 1996). The same research also found that 84 per cent of respondents felt the level of racism had declined over the last five years. Nevertheless racism is still a problem at football matches and needs to be addressed. Football is not the only sport that suffers from racism. At the 1996 Test Match between England and Pakistan at Headingley, there was serious chanting of racist abuse by some 'England fans' (*The Observer*, 18 August 1996). The authorities at sporting venues have a legal, as well as a moral, obligation to prevent fans from chanting racist abuse as it constitutes a criminal offence.

Crowd problems

Any event that attracts large numbers of people faces potential crowd problems. There have been a number of tragedies at sporting events and pop concerts in recent years where fans have lost their lives. The most prominent example was the Hillsborough disaster of April 1989 when 95 people were crushed to death at the Liverpool versus Nottingham Forest FA Cup semi-final. In the previous year at the Monsters of Rock concert at Donnington two fans were crushed to death during a crowd surge. Fire is another risk at events where there is combustible material, as occurred when the stadium caught fire at Bradford City football ground in 1985 and 56 died. Often at special events there are temporary structures and there is a risk of these collapsing. In Corsica in the 1992 European Cup quarter-final between Bastia and Marseilles a temporary stand containing 3,000 spectators collapsed, killing 11 and injuring 863. A smaller collapse of temporary seating occurred at a Pink Floyd concert in October 1994, but luckily nobody was killed and only 36 injured. The very nature of special events also means there are more health and safety hazards such as electrical cables exposed, equipment that could fall down and temporary structures that might collapse.

Some of these risks are more health and safety related than security concerns. The security firm supplying the stewards to the special event will often have a responsibility for crowd safety and it is therefore an issue they need to understand. In some cases the problems with crowds can emanate from security risks. For example, overcrowding can often be caused by people getting in with counterfeit tickets or gaining entry by climbing over fences. The potential for crowd problems and the very grave consequences should things go wrong make this a very important issue to address, not to mention the loss of revenue that could also occur.

Drugs

Drugs are a concern associated with a number of forms of entertainment. Pop concerts, clubs and more recently raves are often associated with drug taking by young people. The use of illegal drugs poses problems for the owners and managers of entertainment establishments, as the discovery of them could lead to prosecution and, in the case of licensed premises, the loss of a licence. The health problems associated with drug abuse, however, are of far greater importance. With the growth in use of Ecstasy (MDMA) at raves in recent years there have been a number of deaths, most notably the teenager Leah Betts. Moreover, it was estimated there were 14 deaths attributed to MDMA in 1991 (Home Office, 1994d). Arguably the use of the drug has grown since this period and therefore it is reasonable to assume that the level of deaths from its use has also increased.

There is also a long list of athletes, pop stars, actors and other entertainers who have taken drugs. The exposure of Ben Johnson, in the 1988 Seoul Olympics, as a drug taker, and Maradonna's positive drugs test and consequent early trip home, in the 1994 football World Cup, illustrates the international scale of the problem. In Britain the problem was highlighted by Paul Merson's public admission of his alcohol, cocaine and gambling addiction while playing for Arsenal football club. The former England football captain, Tony Adams, has also struggled to overcome his addiction to alcohol. Clearly drugs are a major concern throughout the entertainment industry and elsewhere.

Fraud

Frauds are widespread in the sporting world. The most common allegations of fraud occur in sports associated with gambling such as horse and greyhound racing. There have been numerous allegations in the past regarding the doping of horses and greyhounds as well as jockeys being 'nobbled'. In football there have been a series of allegations against a number of British footballers over match fixing related to Far East gambling on premiership matches. There have also been widespread allegations that 'bungs' of money are given to football managers and agents to 'smooth' transfers.

'Excessive' security staff

Security staff themselves can also be a risk to the safety of the public. Poorly trained personnel and incompetent security at an event where there are large crowds, also presents a serious risk. The security employed must be appropriate for the particular event. For instance, at a boxing match where the crowd can get very rowdy, excited and often violent, the security staff need to be capable of dealing with scuffles and are perhaps more zealous than security at other events. Difficulties could emerge, however, if the same type of security was used at a pop concert for teenagers. One could also imagine the potential problems that might emerge if inadequately trained security staff were used at a pop concert in the pit area in front of stage. Many fans could be injured or even killed if the security staff failed to manage the crowd properly in this pit area.

Terrorism

Entertainment has not escaped the interest of terrorists. The nature of many sports and entertainment events, with the very high profile media coverage they attract, often on a world scale, has led many terrorists to realise their scope for attack. At the 1996 Atlanta Olympics a pipe bomb exploded during a free concert at the Centennial Olympic Park killing one and injuring many. At the 1972 Munich Olympics, probably the worst terrorist attack ever at a sporting event occurred when 'Black September' terrorists took nine Israeli athletes hostage after having already killing two. All the remaining hostages were killed as well as five terrorists and a policeman in a bungled police raid. The 1997 Grand National was also postponed due to a PIRA coded bomb threat.

Other risks

There are a wide range of other risks that affect different forms of entertainment. Ticket touts are a concern in most forms of entertainment where the demand for tickets exceeds supply. The problem is particularly acute in some sports, such as football, given the risk of hooliganism. Touts often sell tickets to supporters from one club in areas that are designated for the fans of the opposing team. This makes it harder for clubs to segregate opposing fans and increases the risk of trouble within the stadium. Touts selling counterfeit tickets also heighten the risk of crowd congestion, not to mention the defrauding of the venue and the spectator. A risk peculiar to nudist festivals is 'peeping toms'. At the annual Nudestock at Turtle Lake in Michigan, USA, where a pop festival is held with naked entertainers and spectators, there is a problem of 'peeping toms' attempting to view the entertainment and crowd.

At theme parks and other locations attracting large numbers of children, there are often problems of children becoming separated from their parents or guardians. Such locations may also be tempting for paedophiles and is yet another risk for the security practitioner to address. There is the risk of people being knocked down when they venture into vulnerable areas at sporting events, such as rallies and motorcycle races, where there are barriers (which may also be temporary) to keep spectators off the track. Again this may not be a security risk but, as stewards and security officers are often responsible for keeping spectators in safe areas, this is yet another challenge to overcome. Different forms of entertainment also face the more ordinary security risks associated with arson, burglary, robbery, theft and violence. Many of the strategies that will be used to combat these are the same as those in other sectors, such as retail security, which was discussed in chapter 13.

The origin of risks

The previous section illustrated the wide range of individuals and organisations that pose risks to those in the entertainment industry. Even ordinary fans can be caught up in the hysteria and atmosphere of an event, which can lead to crowd congestion. The obsessive fan can also become a risk by attempting to reach their idol or even attempting to kill him/her. Petty criminals can also prove to be a problem by pursuing crime at entertainment events. More organised criminals may also pose a risk with regard to frauds and other criminal scams. Protesters and terrorist groups also often target certain entertainment events to further their causes.

Strategies to address risks

A wide range of strategies to address the various risks in entertainment security have been developed. The first type can be broadly categorised as national strategies. This encompasses the licensing and regulation of entertainment, national guidelines and codes of practice issued by government, trade, and sporting bodies, and the creation of offences. The second type of strategy includes the actual strategies used by venues and promoters. This includes the employment of security/stewarding personnel, the use of security products, the development of certain procedures and the involvement of public organisations such as the police and local authorities (see Frosdick and Walley, 1997 for a detailed study of sport and safety management). The following section will describe these strategies, their development and how they are used.

National strategies

1. Licensing and regulatory measures

There is no statutory regulation of entertainment security, but there is statutory regulation of certain sporting venues and public entertainment. In many cases there are conditions attached to the main requirements relevant to security. In 1975 the *Safety of Sports*

Grounds Act established a statutory licensing system for specified grounds. These regulations required football grounds designated by the Secretary of State (the old First and Second Division and Scottish Premier) to gain a safety certificate issued by local authorities in consultation with the police and building authorities. When imposing conditions on the issue of the safety certificate the local authority usually refers to the *Guide to Safety at Sports Grounds*, more commonly known as the 'Green Guide', which includes sections on stewarding and crowd behaviour. Since the 1975 Act the 'Green Guide' has been updated on a number of occasions. Designated status has been extended to lower football divisions and to certain grounds in other sports such as cricket, rugby union and league. There have also been a number of regulatory changes extending designated status. The 1987 *Fire Safety and Safety of Places of Sport Act* gave the Secretary of State the power to designate any ground requiring a safety certificate regardless of capacity. The 1989 *Football Spectators Act* established the Football Licensing Authority to oversee the regulatory activities of local authorities, amongst other functions (de Quidt, 1997).

There are numerous statutes regulating entertainment, the most important of which are the 1982 *Local Government (Miscellaneous Provisions) Act*, the 1963 *London Government Act*, the 1982 *Civic Government (Scotland) Act*, the 1964 *Licensing Act* as amended by the 1988 *Licensing Act*, the 1976 *Licensing (Scotland) Act* and the 1967 *Private Places of Entertainment (Licensing) Act*. The conditions that are attached to licences are varied but they frequently include stewarding. In a typical example, the Occasional Licence for Public Weekday Music and Sunday Music set conditions regarding stewarding for the Madonna concerts at Wembley stadium on 25 and 26 September 1993. Conditions included the age of the stewards and the minimum number required, coupled with guarantees regarding adequately trained personnel. The *Guide to Health, Safety and Welfare at Pop Concerts and Similar Events*, which is known as the 'Purple Guide', is used as a basis for setting conditions on licences and includes a section on stewarding.

2. Other national standards/guidelines/forums

Most sports have governing bodies that not only set rules for their sport, but also deal with other issues, which often include security. These bodies may also provide forums for the exchange of ideas on best practice and the sharing of intelligence. The government produces important guidelines on safety at sports grounds and pop concerts in the form of the 'Green Guide' and 'Purple Guide'. The Football League, Football Association and FA Premier League, in consultation with the Football Licensing Authority and Football Safety Officers Association, have produced guidelines on the *Stewarding and Safety Management at Football Grounds*. The guidelines cover issues such as safety policy, safety audit, steward numbers, appointment, training, duties and appearance. The Test and Country Cricket Board has not produced a similar document, but it does hold occasional meetings for chief stewards at cricket grounds to share best practice and intelligence. Other sporting bodies provide similar forums. Some non-sporting/entertainment bodies also produce guidelines that touch upon security at these events such as the BSIA's *Guidelines for the Surveying, Planning and Operation of Stewarding Services In Stadia and Sporting Venues*. This includes sections on stewarding and planning, staffing and stewarding services and operations. SITO has also been developing national standards of training for stewards and supervisors, although at the time of writing they were still to be finalised.

3. The creation of offences and additional powers for the police

The changing nature of entertainment and the way some security risks have developed have led governments to resort to classical methods in addressing certain forms of social behaviour by creating new offences and penalties for committing them. Clearly there are a wide number of crimes, such as theft, assault, robbery and murder, which are already covered by the law. It is new offences and the powers that the authorities have been granted to deal with them, which are of interest. Some of the more prominent examples include the 1986 *Public Order Act,* which created the offence of disorderly conduct to deal with troublemakers at football matches. The 1991 *Football (Offences) Act* created a number of new offences at football matches, which included the throwing of missiles on to the pitch, the chanting of obscene and racist abuse, and invading the pitch without reasonable excuse. The 1994 *Criminal Justice and Public Order Act* created criminal offences of the selling of tickets on match day without authority from the club to do so, as well as giving powers to the police to prevent raves. These are just a few examples of how the state may attempt to address certain risks.

4. National policing

A number of policing bodies are also engaged at a national level in entertainment security in its broadest sense. The National Criminal Intelligence Service (NCIS) collates intelligence on football hooligans, which is disseminated to police forces. There is also an ACPO Public Order Subcommittee that takes an interest in this area and develops policy. Where there is a terrorist risk there is also the anti-terrorist policing infrastructure of the Security Service, Anti-Terrorist Squad and Special Branch.

Local strategies

1. The public forces

The police often play a role in security at certain entertainment events. At football matches a few years ago they would often be the primary force, although their expense has led many football clubs to reduce the number of police officers required by hiring more stewards instead. Nevertheless, police officers can still be found inside the stadium at many football matches and other sporting events and places of entertainment. Police officers inside a place of entertainment are usually there to secure public order and to deal with the more serious problems. Police are also often located outside a large event in order to manage traffic, deal with ticket touts, as well as maintain law and order. Trading standards officers may also be at events, both inside and outside, searching for those selling counterfeit and unauthorised goods.

2. The employment of security/stewarding staff

Often the most important and most common security strategy in the entertainment sector is the employment of security staff to undertake a variety of roles. These usually include: access control, conducting searches, enforcing organisational rules and the law, maintaining order and crowd management to name the most common. The security staff may include security officers, either in-house or contract; stewards employed directly or supplied by a security firm or recruitment agency; and door supervisors again either contract or in-house. Some venues, such as the National Exhibition Centre in Birmingham, have their own in-house security officers, whilst others, such as the South Bank Concert Hall, have contract security officers from companies such as Group 4. Stewards are often employed in-house by a venue, such as

Walsall Football Club, supplied by a security company, such as Showsec, or even an employment agency such as Recruit. Some in-house stewards are unique, such as Twickenham's Honorary Stewards who form around half the 1,000 plus personnel used on a match day. These stewards consist of senior members of the police and armed forces, company directors, surgeons, a priest and even a member of the House of Lords! The changing nature of events and security risks at venues often means that varying combinations of different security staff will be used.

As well as the general security and stewarding staff there are often specialist security staff employed. The more famous sporting and entertainment celebrities often employ their own close protection officers to protect them from over zealous fans, as well as the more dangerous fans. Often sporting and entertainment events have corporate hospitality or other forms of exclusive location. These are often guarded by a Commissionaire, separate from the main security force, who has a security role in ensuring that only those accredited are allowed admission.

3. The use of security products

Numerous security products are also used to ensure effective security, although in comparison to other industrial sectors it is a more personnel based sector. One of the most popular technologies used in recent years is CCTV to monitor crowds and specific areas. Temporary events, such as pop festivals, in effect have to build a small town, and part of that temporary structure includes mobile CCTV systems. Another security strategy that is required for temporary events, unless they are free, is the erection of fencing to prevent intrusions. Even at free festivals and at most other festivals there will be exclusive areas for performers and VIPs that are protected by additional fencing. Other security products that are often used at places of entertainment include barriers at pop concerts and related events to help manage the crowd and prevent them from getting on to the stage. Often there are also access control systems, the more common being turnstiles at stadiums. Communication equipment is also essential at most places of entertainment so that security staff can communicate with one another and with their control room.

4. The development of procedures

A range of procedures have also been developed to enhance security at places of entertainment. One of the most important strategies is pre-planning. This will often form part of a wider strategy of planning at an event. Risk analysis of an event is designed to ensure that all the risks identified are reduced or removed. Such analysis is often an integral element of planning meetings convened long before the event is held. These meetings generally involve the venue staff, promoters, police, fire, ambulance, security staff and the local authority. There may also be a meeting of all security staff on the day of the event, or before, to brief them on procedures and risks. On the day of an event there will also usually be some form of control room or someone ultimately in authority. At a small event this might amount to a security supervisor, but at larger events it might include representatives from the emergency services, the venue and security staff. At very large events there may also be a number of control rooms. Other procedures that are used as security strategies include the segregation of the crowd, restrictions on the sale of alcohol, and all seater stadiums.

Conclusion

In the future entertainment security will continue to provide new challenges for the security decision-maker. The global media attention that some events attract may make them attractive targets for terrorists and other protest groups. This has already occurred at the Atlanta Olympics. The risks to entertainers may also increase. In recent years there have been a number of attacks on officials at football and rugby matches and this may represent the beginning of a trend where such officials require greater protection. There has also been a concern that hooliganism is increasing again and there have been some fears that it may spread to other events such as cricket and rugby.

Chapter 16

Ministry of Defence Security

There are many reasons why the Ministry of Defence (MoD) provides an excellent case study in a comparative analysis of the British security industry. Few organisations have a history of policing and security that goes back to the seventeenth century. As the centre of decision-making and administration for the British Armed Forces, the MoD also presides over a budget of approximately £22 billion, one of the largest in Whitehall. The number of directly employed staff in the MoD and its 44 Defence Executive Agencies (over 300,000) also far exceeds that of any other government department. The MoD's equipment procurement expenditure alone, is approximately £10 billion per annum. The majority of this equipment is manufactured in the UK. The MoD supports an industry of some 420,000 workers who, together with the MoD's own workforce, represent some 2.8 per cent of the UK workforce. In 1997 — 98, 10 contractors, including British Aerospace (now known as BEA Systems) and Hunting and Vickers, were paid more than £250 million by the MoD.

In addition to these budgetary resources, the MoD is responsible for guarding the defence estate. This enormous set of structures, both civilian and military, ranges from the main building in Whitehall to regimental headquarters and barracks, naval bases, airfields, hospitals, service housing, training areas, camps, storage and supply depots, research establishments and radio stations. The MoD is one of the largest landowners in the country, owning the freehold in 1998 to 220,000 hectares, the leasehold to over 17,000 hectares and the rights to 124,500 hectares. Its holdings of property, equipment, ships, tanks and aircraft are extensive. In addition, the defence estate houses many historic buildings, museums, forts, arsenals and military academies containing precious works of art.

These vast assets are potentially vulnerable because such a wide range of groups targets them: foreign powers, saboteurs, hackers and protesters. The MoD is guarding personnel, land, equipment and intellectual property. In order to protect these assets the MoD is a massive user of the equipment provided by the private sector, in terms of CCTV, shredders, fences, alarms and access control. The defence of MoD assets has also encouraged a myriad of different security and guarding organisations. Over the years within the MoD a multiplicity of policing, security, regulating and law enforcement authorities have evolved.

The range of security risks to the MoD

The risks to the MoD, its buildings, personnel and equipment are far greater than any other British organisation. The potential targets are extremely varied. The MoD owns several thousand sites encompassing remote air fields, nuclear bases like Faslane, the highly sensitive chemical and biological defence establishment at Porton Down, and DERA.

MoD establishments, personnel and intellectual property were constantly targeted by foreign intelligence services during the Cold War. However, the MoD remains potentially vulnerable to espionage in light of the military and commercial secrets it guards. The end of the Cold War has not quelled Russia's aspirations to acquire commercial and military information.

Furthermore, there is a great deal of research being conducted by British defence contractors that would be the object, not just of foreign intelligence services, but also industrial competitors, even in NATO member states.

Almost every item within the defence establishment has some value either to petty criminals, industrial competitors or terrorists. Official documents, particularly those with a high protective marking or dealing with the design or deployment of modern weaponry, are vulnerable to espionage. The criminal underworld and terrorist organisations have a constant requirement for small arms, ammunition and explosives. Portable surface-to-air weapons are in particular demand.

Terrorism, rather than espionage or protest, is now agreed to be the most worrying threat to the physical security of military installations. International terrorist groups are developing new methods of operation with expanded areas of interest such as information warfare. The proliferation of biological and chemical weapons, which may be used intentionally by states and terrorist organisations, is also viewed as an increasingly serious threat.

In addition to emerging terrorist threats, the British government faces the possible recurrence of violence in Northern Ireland if the Good Friday Agreement fails to establish a lasting peace. It would be premature to assume that the threat posed by various dissident Irish terrorist groups has also been eradicated.

The MoD is the object of much pressure group activity, mostly from individuals and organisations mounting peaceful protests. For example, protests against the deployment of US nuclear weapons on British territory reached an intensity during the 1980s. Peace campaigners protested outside the military bases at Greenham Common, which housed cruise missiles. The Polaris/Trident submarine base at Faslane in Scotland was and still remains a source of protest group activity. These protests can range from demonstration and permanent peace camps, to groups seeking to penetrate perimeter security.

The animal welfare movement contains organisations that pursue normal pressure group methods. However, it also encompasses those that engage in violent action. MoD agencies such as the Defence Animal Centre could be potentially vulnerable to action from militant animal rights organisations. The centre trains military working animals to help in protecting defence installations at home and abroad.

In common with any other large organisation that controls a massive budget the MoD is also susceptible to the danger of fraud and theft. The National Audit Office (NAO), assisted by Touche Ross consultants, examined fraud in the defence procurement process and what should be done to diminish the risk. The need for such a review was reinforced by the convictions in 1993 and 1994 of three former MoD officials. The NAO's review highlighted the fraud of Gordon Foxley, who in the early 1990s was estimated to have received at least £3 million in corrupt payments. The Chief of Defence Procurement and the Second Treasury Officer of Accounts gave evidence to the Public Accounts Committee (PAC), following a report, by the Comptroller and Auditor General on the risk of fraud in defence procurement. Commenting on the PAC's inquiry into the report, Chairman Robert Sheldon stated that the Foxley debacle was one of the worst cases the committee had ever investigated. The Chief of Defence Procurement admitted that Foxley's deceit represented the largest individual case of fraud in government.

From 1985 to the end of 1994, no fewer than 191 cases of alleged procurement fraud were reported to the MoD. The MoD Police estimated that the value of possible fraud under investigation was about £22 million for 1993-94 (PAC, 1995). In its defence the MoD said that levels of fraud were 'low compared with those in industry and commerce'. The MoD also suggested that the amount should be placed in context, given the department's total budget of £22 billion and the annual procurement budget of over £9 billion. Nevertheless, the Foxley case raised the issue of whether the system had failed to detect an isolated case of corruption, or whether his crimes were but the tip of an iceberg. In response to the case the MoD has initiated an extensive risk assessment of what information might be vulnerable on its computer systems, established a fraud unit and laid down stronger rules on accepting gifts.

The NAO and the PAC have also delivered strong warnings about the threat to government departments that are increasingly reliant on information technology for the efficient and effective delivery of public services. This is particularly relevant to the MoD given the enormous repository of information that could become available to adversaries, industrial competitors and skilled hackers if the appropriate security measures were not taken. Constant vigilance is required and the NAO has identified scope for improvement in the measures taken by the MoD in the design, operation and management of computer systems to combat the risk of fraud and other irregularities.

Security decision-making within the MoD

Given the wide range of interests involved, security decision-making within the MoD is far more complex than in any other UK private security business entity. The extensive network of structures with responsibility for security includes committees within the MoD (the most prominent being the Defence Council), and other government departments, boards and advisory groups. These organisations bring together expertise in many different areas of policing, guarding and security.

At the pinnacle of the decision making process is the Prime Minister and the Cabinet. The Secretary of State for Defence is assisted by the Minister of State for the Armed Forces (Minister - AF) and the Minister of State for Defence Procurement (Minister - DP). Each of these ministers has a specific brief covering policing, guarding and security. Minister (AF) is responsible, inter alia, for operations that include: Northern Ireland, military aid to the civil authorities, nuclear accident response and MoD police operations. Minister (DP) is responsible, inter alia, for intelligence and security policy, nuclear procurement (including safety and disposal) and defence industrial issues. The Parliamentary Under-Secretary of State for Defence is responsible for civilian and MoD Police personnel policy and casework. Additional responsibilities encompass the defence estate and works (including service housing, heritage and historic buildings).

Within the Central Staff the Second Permanent Under-Secretary of State and the Vice-Chief of the Defence Staff are the most senior officials with significant responsibility for policing and security. The senior civil servant reporting to them is the Assistant Under-Secretary (Security & Support) (AUSSS). The present incumbent is responsible to ministers, through the Deputy Under-Secretary of State (Civilian Management) and Second Permanent Under-Secretary, for the organisation and maintenance of protective security and counter-terrorism. The AUSSS is also responsible to the Deputy Chief of the Defence Staff (Commitments) for the policy on the protection of MoD personnel and assets against terrorists and other extremists including setting the counter-extremist alert states. The AUSSS is also chairman of the Security Policy Advisory Group and clerk to the MoD Police Committee. The office holder has the heavy responsibility of acting upon intelligence from a variety of internal and external agencies.

The AUSSS is assisted in these tasks by two members of the Senior Civil Service reporting directly to him. The Director of Security Policy formulates and promulgates defence security policy, sets MoD security objectives and provides guidance on their management and resource implications. He/she also contributes to the formulation of government protective security policy and represents the MoD in interdepartmental and international discussions on this policy. Aspects of security policy covered by the Director of Security Policy include nuclear security, physical security, information technology security, information security, counter-intelligence, security of personnel (including vetting policy), and security education and training. The Director of Security Policy also receives advice and intelligence on threats to security from the security service, the Cabinet Office, Home Department Police Forces, the MDP and its own Defence Intelligence Staff. The Director of Security (HQ, PE and Industry) is responsible for security in the MoD's headquarters buildings, including the MoD Guard Service (MGS) guards employed there. He/she is also the Security Authority for the Defence Procurement Agency, other central agencies, the Atomic Weapons Establishments, DERA and List x companies. There are four other MoD Security Authorities (for each of the Armed Services and for the Permanent Joint Headquarters), who are responsible for overseeing the implementation of defence security policy and standards in their areas.

Policing and security strategies

The Ministry of Defence Police

The present MDP was formed in 1971 when the array of Admiralty, Army Department and Air Force constabularies were unified into a single force. The MDP is a national civilian police force whose officers are appointed by the Secretary of State for Defence and are subject to the provisions of the MoD *Police Act* 1987 (Johnston, 1992b). While the Secretary of State has ultimate responsibility for the operation of the MDP, most of his responsibilities are delegated to the Second Permanent Under-Secretary, who is also the chairman of the MoD Police Committee (although the Minister (AF) now chairs one meeting a year). The MDP Agency Management Board is headed by the Chief Constable of the MDP based at Wethersfield.

The MDP's primary role is crime prevention and criminal investigation within the defence estate. In addition to providing a comprehensive police service, the MDP is trained to use weapons and can therefore conduct armed patrols in defence of personnel and property at certain MoD and other establishments. The MDP also provides police services for the military forces of other governments based in this country and for UK departments, including the Royal Mint. It may also be called upon to investigate crime committed by service personnel where, for instance, civilians are involved. The officers of the MDP possess the same legal powers as their colleagues in the Metropolitan and County forces. Their position, however, is different in a number of respects, given the sensitive nature and value of the material held by the MoD and the fact that they are required to carry arms. The MDP is subject to inspection by HM Inspectorate of Constabulary as a non-Home Department police force.

The future of the MDP in the static guarding role appears uncertain as the cost of employing the force to carry out a guarding function is higher than employing service personnel. The MoD regards the use of the MDP for such duties as 'both expensive and inappropriate', when they do not require the employment of police officers. Pressure for reform has increased with the recurrence of ideas such as arming some MGS officers, further contractorisation, greater reliance upon Home Department police forces, and more extensive use of guarding by the Armed Forces or service police. One of the latest manifestations was the pilot study carried out by the army on the establishment of the Military Provost Guard Service (MPGS), which comprises soldiers on Military Local Service Engagements (MLSE). The MPGS concept is intended to relieve 'full engagement regular soldiers' from guard duties, with the aim of improving manning and retention in the Field Army, and to replace MDP officers employed on guarding duties that do not require constabulary powers. The MPGS concept was highly criticised at the outset by the House of Commons Defence Committee (HCDC) and by the Defence Police Federation, as it was seen as a further measure to downsize the MDP, but it is now, after a feasibility study, being implemented.

With further cuts envisaged questions have been raised about the minimum size of viability for the MDP. The MoD considered that a force of 2,500 to 3,000 would be viable, if properly structured and resourced. This estimate was formulated on the basis of the tasks that need to be undertaken by police officers. However, the Defence Police Federation told the HCDC that the proposed cuts were 'so drastic as to endanger the effectiveness, viability and resilience of the force'. They argued that comparisons between the size of the MDP and civilian police forces were invalid because the latter were located in one area, whereas the former are often stationed in small detachments throughout the UK. These misgivings were shared by the HCDC. The committee has expressed the view that a MDP of such size might be not be able to meet all unforeseen emergencies, particularly if they were for a prolonged period. At the end of March 1999, the MDP complement was 3,741, with an actual strength of 3,589 but, on current assumptions, the size of the force could fall to between 2,500 to 3,000 personnel in the early years of this century.

The Ministry of Defence Guard Service

The MGS was formed in October 1992 from a variety of patrolling and watching grades, both industrial and non-industrial. Prior to the formation of the MGS, civilian guards at MoD establishments were provided through directly employed labour and contractors. The three services and other MoD organisations trained their own patrolling and watching grades to varying standards. Following an upsurge in PIRA activity in the UK in 1988-1989, which culminated in the attack on the Royal Marines at Deal in September 1989, a MoD study made various recommendations regarding the need for a unified MGS. The objectives established for the new service included the rationalisation of guarding arrangements and the improvement of standards. Perhaps the most important objective was freeing service personnel and the MDP from routine unarmed guarding duties and from posts with no requirement to exercise constabulary powers.

In essence, the MGS serves as the MoD's in-house security force, with the Chief Constable of the MDP responsible for its professional standards and training. The MGS is committed to provide the MoD with a high quality and professional service, which is alert and well-trained, well-informed on possible threats and capable of integration with other guarding forces. Accordingly, the MGS has a uniform, professional standards, grading structure, common pay, and conditions of service. Its personnel are carefully selected and, by the standards of the private security industry, are very well trained. The MGS has not been armed, but some guards have been trained as dog handlers. It conducts all unarmed guarding tasks such as access control, patrols, searching, and the issuing of passes and permits to visitors. MGS personnel have the standard terms and conditions available to non-industrial civil servants.

Despite some initial scepticism the establishment and operation of the MGS has been an enormous success, though the private sector is nibbling away at the edges and it is being subject to increasing competition. The MoD has expressed support for the MGS but there is pressure for a reduction in its size. In 1996 the MoD set a cost reduction for the MGS of 12 per cent. The HCDC concluded that 'very serious consideration should be given before any contractorisation or market resting be proceeded with' (HCDC, 1996). Conservative government ministers stated that if the savings envisaged by the department were not realised, then it would consider a policy of limited market testing and contractorisation. On 1996 assumptions the MoD expects the size of the MGS to fall from its current strength of 4,400 to some 3,250 to 3,750.

Service police

These are the specialist police forces of the Armed Forces. The Royal Military Police (Army), the Royal Navy (RN) Regulating Branch, the Royal Marines (RM) Police and the RAF Police are each integral to their own service. They exercise jurisdiction over service personnel wherever they are serving, under the *Service Discipline Acts*; and when abroad their jurisdiction extends to all members of the force, including dependants and civilians. The service police support their respective services in Northern Ireland and in HM ships operating outside UK territorial waters. They operate with British troops abroad and undertake the complete spectrum of policing activities including criminal investigation, close protection and drugs investigations.

There are 357 personnel in the RN Regulating Branch, which is headed by a Provost Marshal. The Royal Marines Police are a subdivision of the Royal Military Police. Neither the RN Regulating Branch nor the Royal Marines Police have any security responsibilities, although individual members may be appointed to security/guarding posts. The RN has area security teams offering advice and conducting the mandatory security inspections of all RN/RM establishments. Naval dockyard security is headed by a base security officer. At RN/RM manned units service personnel undertake the majority of the armed guarding task. At RN units the MGS undertake the majority of unarmed tasks whilst at RM units contract guards undertake the majority of them.

Security at the Faslane submarine base is probably the best in the country as it is primarily the home base of the Trident (nuclear submarine) flotilla. The MDP, the MGS, the RM and the Strathclyde police coupled with the best security equipment available to the MoD are all deployed to provide high quality security at a particularly sensitive location.

The Royal Military Police, which has a strength of about 2,000, is headed by a Provost Marshal at Upavon. He has additional responsibilities for the professional management and training of the Military Provost Guard Service. Soldiers are provided for both armed and unarmed tasks on an uncomplemented basis when resources permit. Units within a smaller military population provide an armed guard when possible and the unarmed work is the responsibility of the MGS.

There are 2,321 RAF Police and 146 Provost Officers headed by an Air Officer Security and Provost Marshal. They perform a combined security and policing role that encompasses the full spectrum of counter-intelligence, protective security and police duties on behalf of RAF and NATO formations. The RAF is presently examining internal proposals to replace a proportion of the RAF Police with another form of manpower. The latter would conduct the majority of the armed guarding tasks but would be supplemented by other RAF ground tradesmen diverted from their primary tasks, for what it would be hoped would be a maximum of 1 week in 26.

As any ex-servicemen will testify, a spell of guarding is usually an unwelcome addition to their more regular military duties. Each of the services may differ slightly in their use and numbers but armed guarding remains essential

Military Provost Guard Service

The objective behind the formation of the MPGS was to replace the MDP with soldiers recruited on local service engagements. According to the MoD the MPGS would only be replacing the MDP 'in its armed guarding role'. The MPGS was launched by a Defence Council instruction in 1996 as 'a non-police adjunct of the Adjutant General's Corps (Provost)' from the beginning of April 1997. The objective of the new service was to rationalise existing guarding arrangements (normally at HQ LAND Anti-Terrorist Security Measure Category 'A' sites where soldiers live and work) mainly by relieving MDP officers assigned for armed guarding and general security duties.

The command and control of MPGS soldiers will lie with the Commanding Officer or Head of Establishment at the sites to be guarded. Personnel will be generally recruited from ex-service personnel who have left the Armed Forces, although in certain circumstances applicants with no prior military service will be considered. Proper evaluation of the MPGS can only be conducted when the MoD makes available its report on the outcome of the pilot scheme.

Contract security firms

Contract security firms have been employed within the MoD and by the services for nearly a quarter of a century. Functions of private security on designated MoD sites involve: access control, gate duty, some patrolling, traffic management, fire prevention and the searching of vehicles. Over the last 10 years there has been a noticeable increase in the use made by the MoD and the services of commercial security guards. This is evident not only in the number and value of contracts, but also in the nature of the locations at which they are employed. In real terms expenditure on private security contracts rose almost tenfold between 1984-85 to 1989-90.

Despite the then existing array of policing and guarding forces at its disposal, the private security industry appeared most attractive to the MoD as a supplement. The particular attraction was cost. The unregulated private security industry could be employed at much lower cost, bearing in mind the low wage rates, poor training and supervision. In addition to expenditure considerations the use of private security firms allowed the release of highly trained servicemen to matters of greater military priority, thus enabling them to use the skills that they alone have. A further factor was the Thatcher government's privatisation programme and its desire to reduce the MDP's numbers. However, the deficiencies of the private security industry became apparent particularly through the continuous interest displayed by the HCDC, of which Bruce George has been a member since 1979. The HCDC has produced several reports on the security of military installations over the past two decades, which have highlighted concerns at the use of contract security by the MoD (HCDC, 1984, 1990 and 1996).

These reports have led to the MoD attempting to improve the standards of private security that it uses. Initially the MoD issued guidelines to their contract branches on the hiring of commercial guarding companies. In October 1995 more stringent standards were imposed as the MoD issued *Security Instructions For The Use Of Unarmed Commercial Guard Forces in Great Britain*. Known informally as the 'Yellow Book', the instructions are made available to prospective clients as well as security directors. This document advises hirers of the 'minimum acceptable security and employment standards required from the commercial company delivering the contracted guard force'. These highly detailed instructions are not only designed to ensure that a competent firm is selected, but also to assist with the monitoring process. Contract guards are used as the sole guard forces at those sites considered to be less attractive to criminal or terrorist organisations. They may be used at more sensitive sites when integrated with a directly employed guard force, such as the MDP or MPGS (if there is an arming requirement), or MGS (where there is not).

Even the much improved new system is not above criticism. Security provided by a contract with the MoD is of a completely different nature to that of the MDP, the MGS and the service police. Commercial security guards who take responsibility for security at certain sites undergo a three week training course. Many guards also study for City and Guilds qualifications and NVQs. Nevertheless, they are not subject to rigorous training courses as are MoD forces, nor do they undergo the same degree of screening. Consequently, commercial security guards are unarmed and have no constabulary powers or jurisdiction outside of MoD property. Certainly, stricter regulation has led to considerably improved performance by approved companies. Other pressures, however, have also encouraged firms to improve their performance with better salaries, training and education. The introduction of the minimum wage, EU directives on the limitation on hours and, it is to be hoped, statutory regulation should all help to reinforce these trends. Following a quality review of contract guarding conducted last year, action is in hand to improve the regulation and monitoring of training standards under the SITO umbrella.

List x companies

Any company that is sponsored by the MoD to undertake work on a contract involving protectively marked material at 'Confidential' or above must meet stringent security standards, which include satisfying the MoD that its guarding arrangements will be adequate to protect defence property and information. The company must satisfy the MoD that it has the correct procedures in place, and must appoint a security controller, designate a board member responsible for security and submit all staff who are to work on protectively marked material for security clearance. The MoD then monitors the activity of the company and its compliance with those standards. Guidance is regularly issued to companies in the form of List x notices, which includes guidance on guarding. Indeed, recently a notice was issued specifically on this subject.

The MoD's view is that where List x companies seek to protect their premises with a commercial guard force, then its security standards must be adequate to protect defence property and information. Companies are required to seek confirmation that the guarding company concerned is registered on the Department of Trade and Industry's Registrar of Quality Assured Companies. Registration on the list means that the company meets both the relevant British and International Standards required and also that it belongs to recognised trade associations, which conduct regular audits by third parties to ensure standards are maintained. If companies are not registered on the list, it is still possible for them to be employed to guard List x premises. Such companies, however, must meet the same criteria as those who are registered. MoD rules also require the guards employed on the contract to hold a personal security clearance in accordance with the material they will be protecting. Whatever the case, the company security controller remains responsible for the security of the site and therefore must be fully satisfied that the guarding company has the will and ability to afford its assets adequate protection.

Conclusion

For historic reasons the British military has guarded itself and its property. For a variety of reasons policing and guarding have become more complex with different structures, and pressure to contract out and cut costs. An additional pressure is the need to free service personnel for military tasks that cannot be performed by anyone else. A counter to this pressure of economy is the trend of specialisation and, in particular, the imperative to respond adequately to the growing threat of terrorism. The earlier criticism of bodies like the HCDC as well as the desire to improve the quality of the limited number of commercial guarding companies, has led to a much more sophisticated system of awarding and withdrawing contracts. There is still, however, the criticism, perhaps unfairly levied, that price continues to be a dominant consideration in the awarding of contracts to the private sector. The private security industry remains subject to heavy criticism given the performance of the commercial sector on some sites. Nevertheless, in light of current financial imperatives the continued use of such companies is still inevitable within well-defined parameters.

Part 4: Regulations and
the Private Security Industry

Chapter 17

Regulation and the Private Security Industry

Regulation has been one of the central issues of importance to the private security industry in recent years. This chapter will start by examining what is meant by 'regulation', as it is a concept that is often confused. It will then move on to explore some of the many ways the government already intervenes in the private security industry to set standards. This is followed by a brief examination of the arguments for and against regulation. How private security has been considered as a policy issue, particularly vis-à-vis regulation, in recent years follows on. The chapter will then explore the increasing European level influence and finally it will speculate on some of the challenges and trends that might confront policy-makers in the future.

What is regulation?

It is important to define what is meant by regulation as it is a term that is often misunderstood, as are related terms such as self-regulation, licensing and registration. At a conference held in 1995 to discuss a proposed regulatory scheme for the private security industry, in his opening address, the chairman argued that the industry required strong independent regulation, along the same lines as the British Medical Association (BMA) and the Law Society. However, the BMA is a professional association and not the regulatory body for doctors, and the General Medical Council, which is the regulatory body for doctors, could be described as the model of statutory self-regulation. The Law Society is the regulatory body for solicitors, but it is also a model of self-regulation.

Many argue for statutory regulation of the private security industry when the industry is already subject to statutory regulations on employment practices, taxation and health and safety to name a few. What the industry is not generally subject to are specific regulations relating to the establishment and operation of a private security firm. Regulation is therefore a broad concept encompassing any state laws, regulations or stipulations of bodies with statutory force, which affect individuals and organisations. According to Francis (1993: 5) regulation can be defined as:

> state intervention in private spheres of activity to realise public purposes.

More specifically Eisner (1993: xiii) defined regulation as encompassing:

> *A broad array of* policies governing economic activity and its consequences. Regulatory policies address firms' entry into and exit from particular markets; prices; rates of return; and modes of competition; as well as the characteristics of the goods being produced, the quantities being produced, the means of production, and the negative externalities (such as pollution) arising from the production process [authors' emphasis].

Some form of regulation impinges on an individual almost every day. During a journey to work a motorist complies with a number of regulations: possession of a driving licence and insurance; maintaining the car to meet necessary safety standards; and observing certain regulations. A motorist switching on the radio will listen to stations that are subject to regulations on which frequency they can use, how much advertising they are allowed to broadcast and what subjects they can cover. The motorist will then arrive at work for an organisation that is subject to many regulations on how it operates.

When most people call for statutory regulation of the private security industry they usually envisage some form of statutory licensing or registration. Both licensing and registration are forms of regulation. Licensing involves authorisation from a body, usually a state body or an organisation with statutory force, before an individual or organisation can lawfully undertake a specific activity or a product can be used. Registration generally involves the submission of information about an individual, product or organisation to a body after the commencement of an activity or the release of a product, although it may also occur before. The standards for registration are usually perceived to be lower than for licensing, particularly in the USA (Kakalik and Wildhorn, 1971c), although in the British context the difference is more one of semantics. Thus, for ease of understanding the term regulation has been used as a generic term throughout this book to embrace the terms licensing and registration, as well as pure regulation.

A number of other terms that are used in this debate also need clarification. Regulation can be divided into external and self-regulation (Swann, 1989). External regulation (independent regulation is also used in this context) is where control is exercised from outside the regulated group by a government, local authority or some other independent body. Self-regulation involves parties who regulate themselves, either through an association or directly. An alternative definition of self-regulation has been provided by Black (1996: 26):

> The disciplining of one's own conduct by oneself, regulation tailored to the circumstances of particular firms, and regulation by a collective group of the conduct of its members or others.

These definitions should now make the history of the regulation debate and the case for and against easier to understand. Before we embark upon this, however, it would be useful to explore government intervention in the private security industry as there are a surprising number of ways in which this already takes place.

Government intervention in the private security industry

Although there has been no general regulation of the private security industry in Great Britain (although as chapters 6 to 10 illustrated there is a complex mix of voluntary, external and self-regulatory measures) there are a wide range of ways in which the government already intervenes in the private security industry. For a detailed study of this you are referred to Button and George, forthcoming. First, as a major, if not the biggest, user of private security services and products, it can influence standards. Second, there are a wide range of specialist areas where statutory standards can be set. Finally, a wide range of guides are also issued that influence private security organisations in their standards. Hence even without the introduction of a statutory licensing system there are a wide range of means open to a government to influence standards. Some of these will now be briefly explored.

We estimate, based upon questions to ministers, Agency Chief Executives and other government bodies, that over £300 million is spent on private security each year by government, in the UK (Button and George, forthcoming). There is no central purchasing, rather it is fragmented to each department and agency so this weakens that power. However, there are guides and evaluations on the purchase of private security that do influence the industry. Most products are evaluated and placed in the 'restricted' document, the *Catalogue of Security Equipment*. Companies meeting these standards and being used by government departments also use compliance as a marketing strategy. Use by the government, particularly a prestigious or sensitive part of it, is also often used as a positive marketing endorsement when attempting to gain business in the private sector.

There are also a number of specific areas where the government already intervenes in the private security industry by setting statutory standards. Private prisons and prison escorts are subject to tough statutory minimum standards (see chapter 7). In the field of transport security — ultimately deriving from international Conventions — there are legislation and regulations mandating minimum standards for aviation and maritime security (see chapter 14). Through other international Conventions there are also regulations mandating minimum standards for civil nuclear installations. Less direct statutory intervention also sets minimum standards of security through health and safety regulations. For instance, legislation requires organisations to protect staff from reasonably foreseeable incidents of violence.

Standards are also set without statutory intervention through contracts and guidelines (as with the MoD, see chapter 16). In other areas of activity government bodies have issued extensive guides on standards of security. These areas include hospitals, schools, courts, and football stadia to name a few. Thus there is already a wide range of means open to a government to influence the standards of private security industry. With the introduction of statutory regulation this will be further enhanced.

The British debate over regulating private security

Before the handling of private security as a policy issue is examined it would be useful to set out the main arguments that have been used for and against regulation in recent years. Probably the main argument advocated for the introduction of a statutory system of regulation has been the inability to prevent criminals working in the industry (George and Button, 1995; and HAC, 1995). It has been argued that those employed in the private security industry undertake responsible and often sensitive roles and it is therefore essential that they are of good character. Advocates of regulation have provided a wide range of anecdotal evidence of security employees with extensive criminal records operating in the industry. A typical example was the 'Cook Report', which set up a bogus security firm of which 48 of the 71 applicants had criminal records (Broadcast, 30th May 1995, ITV)! Evidence illustrating the extent of criminal activities has also been provided by ACPO, which published research illustrating over 2,500 offences committed annually by private security staff (ACPO, 1995).

The second strand to the arguments put forward for regulation has been the poor standards of performance of the private security industry in general. It has been argued that private security is not like any ordinary product or service as its effectiveness has an impact on safety and crime prevention (Loader, 1997). Take, for instance, a personal attack alarm: if this fails or is difficult to operate it might result in a woman being raped or even murdered. Therefore advocates have argued there should be minimum standards of operation to raise standards overall. Some of the evidence that has been provided has illustrated poor rates of pay, long hours worked, low levels of training, intruder alarms that do not work, amongst many others (see HCDC, 1990; George and Button, 1995; and HAC, 1995).

The final strand of the arguments for regulation, which has been the least debated and explored, is the lack of accountability of the industry. It is argued that the private security industry undertakes a range of important tasks, some of them equivalent to the police (Jones and Newburn, 1998; and Button, 1999), for which there are no special mechanisms of accountability. It is argued that suitable structures should be developed to address this anomaly.

Moving now to the arguments that have been used against regulation, the more recent opposition has centred on the free market ideology of the last Conservative government. An integral element of this policy was to pursue deregulation wherever possible, and where there were calls for statutory intervention the government encouraged voluntary measures instead. It was argued that the choice of security products and services was a matter for the buyer, and if they selected those provided by firms beyond the voluntary structure that was their risk. Linked to this was a belief that regulation would only increase the burdens on business and ultimately increase costs (Home Office, 1991; and Murray 1996). Some advocates of the status quo also argued that the evidence of criminality and poor standards was largely anecdotal and lacked clear proof.

During the late 1960s and early 1970s the arguments against regulation followed a very different line. Initially, there was a concern that regulation might give the industry a legitimacy and authority it did not deserve. Proposals for the regulation of private investigators were rejected on the grounds it might give them a 'licence to pry' (*Hansard*, 13 July 1973: col 1966). Similarly, in 1981 Lord Willis's Security Officers Control Bill was opposed by some Lords on the grounds it might give the public the impression they have 'special powers' like those of the police (*House of Lords Report*, 3 December 1981: col 1166). Partly linked to these arguments has been a fear amongst some on the left of the Labour Party that the greater legitimacy regulation might bring may enable further privatisation of the criminal justice system. Now the main arguments for and against have been described, the treatment of private security as a policy issue can now be examined.

Private security as a policy issue

Private security has become a subject for policy-makers for a range of reasons. Most frequently it has emerged because of demands from various interests for some form of statutory regulation or the setting of statutory minimum standards. It has also emerged in debates related to the privatisation of various aspects of the criminal justice system. This section will consider some of the most important occasions where private security has emerged as a policy issue.

Government interest

The origins of policy are diverse but, whatever its antecedents, proposed legislation must generally go through similar stages in the government machine. Generally a Green or White Paper will be issued, depending upon how firm the government's proposals are, although this does not always occur. The government's proposals are then discussed and comment is invited from interested parties. However, alternative methods can be pursued. A department may also organise an internal working group consisting of representatives from certain interest groups and officials to review a policy and make recommendations. A consultation may also be pursued, which even involves comments on draft legislation, as is likely to happen with the Labour government's proposals at the time of writing (August 2000). Legislation may also be introduced immediately without any consultation or detailed consideration. Civil servants can also be given the task of investigating an issue and making recommendations. For instance, in December 1993 ministers asked officials to conduct a review of police tasks. Their report, the *Review of Police Core and Ancillary Tasks*, was published in 1995 and made a number of recommendations concerning the responsibilities that should be relinquished by the police.

Governments have also issued a number of Green and White Papers, as well as organising internal working groups, of relevance to the private security industry; for example on regulation of the industry in 1979, 1991, 1996 and 1999 (Home Office, 1979, 1991, 1996a, and 1999). There have also been papers related to privatisation such as, *Court Escorts, Custody and Security: A Discussion Paper* (Home Office, 1990b). The subject of wheel-clamping and disclosure of criminal records has also been discussed (Home Office 1993a, b; and 1996b). The most significant White Paper so far, however, is *The Government's Proposals for the Regulation of the Private Security Industry in England and Wales* (Home Office, 1999). This at last sets out detailed proposals, which are likely to be implemented and which will briefly be discussed later in this chapter.

Once the decision on a policy proposal has been completed it then has to be implemented. This will usually require legislation or secondary regulations. It will also have to pass the better regulation initiatives (or deregulation initiatives as they were previously). Under the last Conservative government these procedures were a significant hurdle (see Deregulation Task Force, 1996; and Cabinet Office 1996).

There have been a variety of statutes passed that have impacted upon the private security in the last three decades — almost all before the better regulation/deregulation initiatives. They have included legislation mandating minimum standards of security in aviation; statutes regulating an aspect of the industry such as guard dogs; and laws enabling traditional public sector activities to be contracted out to the private sector. Some of the main legislation of importance includes: the *Guard Dogs Act* 1975, *Aviation Security Act* 1982, *Criminal Justice Act* 1991 and *Police Act* 1997 amongst many others. There has also been the introduction of some general legislation that will/ and has had a significant impact upon the private security industry. The *National Minimum Wage Act* 1998 established a minimum wage of £3.60 per hour, which in some sectors, where there has been low pay, has already had a significant impact. Statutory instruments (secondary regulations) may also be issued under some legislation by governments. One such example is the *Working Time Regulations* 1998, which set, amongst many other requirements, a maximum working week of 48 hours (although individuals can opt to work longer). In the static contract guarding sector, where hours are often worked in excess of this, those who do not wish to, do not have to.

Parliamentary interest

Despite general indifference towards private security in parliament over the last 30 years, it has become the subject of increasing attention more recently. There have been a number of attempts to introduce legislation to regulate some aspect of the private security industry through amendments to a government bill. The first was an amendment advocated by Conservative MP, Michael Stern, to the 1994 *Criminal Justice and Public Order Act*, which would have given the Secretary of State powers 'to make regulations for the licensing of the provision of security services'. However, the amendment was never considered because the Speaker ruled it was not relevant to the purpose of the legislation. Stern later introduced a *Ten Minute Rule Bill* in the session that was almost identical to the amendment, which failed.

Three attempts were moved by Labour Home Affairs Spokesman, Alun Michael, during the passage of the 1994 *Police and Magistrates Courts Act* and the 1994 *Criminal Justice and Public Order Act*. All of these amendments were defeated. A further attempt was also made by the Labour Opposition in the House of Lords during the passage of the 1997 *Police Act*. Lord McIntosh of Haringey moved an amendment during the committee stage of the bill that would have required the registration of those providing manned guarding services, supplying and installing security equipment and any other activities prescribed by the Secretary of State (*House of Lords Official Report*, 2 December 1996). This attempt also failed.

The second means by which regulation has been pursued through Parliament is via private members bills. Unless, however, the government supports the bill there is little chance of it succeeding. Given the difficulties faced by a back-bencher seeking to introduce legislation through a private members bill, it is no surprise to find that none have succeeded in regulating the private security industry through this procedure. Nevertheless, there has been no shortage of attempts to regulate the industry through this procedure as the following figure shows. However, none have been introduced through the ballot - which is the most likely means of success for a private members bill. Most have introduced their bills to raise the profile of the subject or in the hope that the government will introduce legislation.

Table 7. Private members' bills introduced seeking to regulate the private security industry

Sponsor	Name of bill and year	Description
Gardner, T (Con)	Private Investigators Bill (1969)	Would have required private investigators to gain a certificate from a County Court Judge, as well as requiring bond (Y).
Walden, B (Lab)	Right of Privacy Bill (1969)	Would have introduced right of privacy, which would have affected operation of private investigators (Z).
Huckfield, L (Lab)	Control of Personal Information Bill (1972)	Forerunner to the *Consumer Credit Act* and could have formed basis for control of private investigators (Y).
Fowler, N (Con)	Security Industry Licensing Bill (1973)	Would have required licensing of all those offering security and investigative services along the lines of the Gaming Board (Y).
Fidler, M (Con)	Private Detectives Control Bill I (1973)	Would have disqualified those with criminal records from describing themselves as private detectives (X).

Sponsor	Name of bill and year	Description
Fidler, M (Con)	Private Detectives Control Bill II (1973)	Would have established a licensing authority for private detectives (X).
George, B (Lab)	Private Security (Registration) Bill (1977)	Would have established a Council to register all those offering security and investigative services (Y).
Lord Willis (Lab)	Security Officers Control Bill (1981)	Would have barred individuals with criminal records from becoming security officers (House of Lords).
Dixon, D (Lab)	Private Security Bill (1987)	Would have required individuals and companies offering security services to register with the Secretary of State (X).
George, B (Lab)	Private Security (Registration) Bill (1988)	Update of 1977 version (Y).
Wheeler, J (Con)	Security Industry Bill (1989)	Would have required firms and individuals supplying security services to register with an inspectorate (X).
	Security Industry Bill (1990)	Identical to 1989 version (X).
George, B (Lab)	Private Security (Registration) Bill (1990)	Update of 1988 version (X).
	Private Security (Registration) Bill (1992)	Update of 1990 version (X).
Stern, M (Con)	Security Industry (Licensing) Bill (1994)	Would have given powers to the Secretary of State to regulate contract guarding security services (Y).
George, B (Lab)	Private Security (Registration) Bill (1994)	Update of 1992 version (X).
Starkey, P (Lab)	Door Supervisors (Registration) Bill (1998)	Would have required the registration of all door supervisors (Y).

Note: Letter at end of description denotes type of Bill: X = Ordinary Presentation;
Y = Ten Minute Rule Bill; and Z = Ballot Bill.

The initiation of legislation is not the only means by which private security has come to the attention of parliament. Opportunities also exist in parliament to influence the decision making process. These can be achieved through debates, questions, Early Day Motions, select committees, party committees and all party committees (Clarke, 1992). These have been used in varying degrees by MPs and Lords to raise the issue of private security. In-particular, the interest of the HCDC must be noted during the 1980s and early 1990s in the absence of any interest amongst other select committees, most notably the HAC (HCDC, 1984, 1990 and 1996).

Other interests

The private security industry has also been the subject of a number of major reports published by various organisations. Outside the direct structure of government and parliament there are a number of means for independent inquiries to be established, which often lead to influential reports. These include Tribunals of Inquiry, Royal Commissions, Departmental Inquiries as well as other inquiries founded independently of the government. There have not been any that specifically focus on the private security industry, although there have been a few that have touched upon it. The most prominent of these was the Younger Committee on Privacy (1972), which investigated whether a right of privacy was required. It also explored the world of private investigators and one of its recommendations was that they should be subject to statutory licensing. Some public organisations also produce reports that make recommendations and influence governments. For instance, the Audit Commission (1996) recently undertook an investigation into the effectiveness of police patrols and made a number of recommendations. A 1988 ACPO leaked report into the private security industry, advocating regulation also proved influential.

Many private organisations produce reports and make recommendations that eventually become government policy. For instance, the Police Foundation and Policy Studies Institute (1996) founded an inquiry into *The Role and Responsibilities of the Police* and published a report with a number of recommendations, one of which was regulation of the private security industry. There has been relatively little interest in the private security industry from think tanks. Proposals by the Adam Smith Institute for the privatisation of certain areas of the criminal justice system provide a notable exception. The 1991 *Criminal Justice Act* and 1994 *Criminal Justice and Public Order Act* introduced and expanded the privatisation of court security, prison escort services and private prisons. Such measures were among a wide range of proposals advocated by the Adam Smith Institute in publications throughout the 1980s and early 1990s (Young 1987; and Elliott 1988, for example).

Private security has also been a subject of interest at an international level. This has led to Treaties requiring domestic legislation affecting private security. Much of the legislation regulating aviation and maritime security originates from such sources (see chapter 14). The prime motivation behind the introduction of the 1990 *Aviation and Maritime Security Act* was to bring into force the 1986 'Rome Convention on the Suppression of Unlawful Acts Against the Safety of Marine Navigation'.

The Labour government's proposals for regulating the private security industry

As the introduction of a regulatory system for the private security industry is likely to have a significant impact it is worth considering the government's White Paper at greater length (Home Office, 1999). As they stand, Jack Straw's proposals represent one of the most significant reforms the industry has ever faced. Although they could go further, they create an efficient and effective structure to regulate the industry and are to be welcomed.

The White Paper proposes a Private Security Industry Authority (PSIA) to licence all employees, managers and directors in regulated sectors. Initially this will focus upon manned security services and alarm installers but there will be scope in the legislation to extend it to other sectors of the industry through secondary regulation. Licensing would be undertaken by the PSIA after obtaining the applicant's criminal record (if applicable) from the Criminal Records Bureau, to be established under the 1997 *Police Act*, which would assess if an applicant is a 'fit person'. The industry would also be exempt from the 1974 *Rehabilitation of Offenders Act,* so any spent convictions could be considered in the application. The White Paper, instead of proposing statutory regulation of companies, proposes a Voluntary Inspected Scheme, whereby existing (and potentially new) voluntary schemes are endorsed and promoted by the PSIA. However, there would be scope to make these provisions compulsory in a particular sector through secondary regulation.

Although the White Paper does not at the moment propose minimum standards of operation, we believe it will have to implement these in the medium to long-term (if not immediately) if it is to achieve one of the fundamental aims of the White Paper: to raise standards. The government is also likely to come under intense pressure from the industry and other important groups to implement minimum standards. These changes, when implemented, are likely to have the following consequences in the longer term. Security companies will have to improve their professionalism in order to meet the new standards that will be imposed upon them. In the current state of the market there will be many companies incapable of undertaking this, which will either go out of business or be closed down by the regulator. Those companies that remain will only do so by raising their standards of professionalism. The higher standards and therefore ultimately higher costs may lead many consumers of security services and products to demand even higher standards for the fees that they pay. This, combined with the changes to be discussed below, will require the standards of professionalism of the employees of the industry to rise even higher. Ultimately the private security industry in the twenty-first century will be slimmer, with fewer companies, but operating to higher standards. As Moore (1995:13) has argued:

> No longer will the security industry be able to pick a person from the streets and say 'guard these premises'. That job will be performed by robots, electronic sensors, and surveillance devices. Private security personnel in the 21st century must have the ability to operate the devices that will perform guard services of today. They must be computer and technologically literate, and conscious of the business, financial, and social climate in which they operate. Private security personnel must be prepared to meet the challenges of the 21st century.

European interest in private security

Much has been written in recent years related to the ideas of the movement towards a 'late-modern', 'high-modern', 'postmodern' or 'risk society' (Beck, 1992; Giddens, 1990; and Johnston, 2000). One of the central tenets of these arguments has been the idea of globalisation. In response to these growing challenges there have been a multiplicity of global initiatives. One of the most important has been European — most significantly European Union — responses to problems and challenges. This dimension has also not escaped initiatives that affect the private security industry.

The European dimension has become increasingly important when considering private security. First, regulations from the European Union (EU) have already been introduced that directly affect the industry, and further measures may be imminent. Second, the creation of a single market means that standards for products and services will increasingly be developed at a European level. Therefore, some existing national standards will merge into European standards.

The issues of policing and security fall within the field of co-operation in the field of justice and home affairs, and the principle of subsidiarity applies. Therefore, any policy in the area of private security is still within the domain of the member states. However, there have been some exceptions in the field of home affairs with regard to the development of European legislation. Council Directive Number 91/477/EEC on the Control of the Acquisition and Possession of Weapons (largely implemented in the UK under the 1992 *Firearms Act (Amendment) Regulations - SI 1992/2823*) provides a notable example. This measure was achieved through the powers to complete the single market. Using these powers it would be conceivable for a Directive to be eventually issued regulating some aspect of the European private security industry. At present there are barriers to CIT crews travelling across borders and contract guarding companies operating in another state. Private investigators who wish to operate in more than one member country also face difficulties. The barriers to a single market in these services might lead to measures at a European level to harmonise standards. Alternatively, measures might be taken to enforce the mutual recognition of standards to facilitate the free movement of private security services and personnel within the EU.

The EU is in the very early stages of policy development in this area. The European Parliament was the first to show an interest following a lobby of European MPs by the GMB/APEX. In February 1991 the European Parliament adopted a resolution calling upon the Commission to put forward legislative proposals for the harmonisation of the European private security industry (APEX, u.d.). The Commission has yet to respond but it has organised a social dialogue for the private security industry (in operation since 1993). The social dialogue has a long history in the European Union and has been used to promote agreements between the social partners and, where appropriate, to consult them on proposed legislation (European Commission, 1996).

The social dialogue for the private security industry was organised by Directorate General V (DG V) (which has since changed to DG Employment and Social Affairs). The social partners invited to attend included Euro-Fiet (a Euro-wide body of trade unions, which represents private security officers, and which has since changed to UNI-Europa), the Confederation of European Security Services (see chapter 5), IPSA and other interested Directorates (DG XV and DG XXII at the time) (European Commission, 1996). So far the social dialogue has resulted in studies into the training of security officers and regulation of the industry across Europe, culminating in the organisation of a conference on the European private security industry in September 1996. Here agreements were signed between COESS and Euro-Fiet on regulation of the private security industry and training. In 1999 another conference was organised, this time in Berlin, on public contracting. However, the European Commission has shown little interest in developing legislation for the industry so far. At the meeting of the social dialogue on 7 March, 1995 a DG XV representative (Internal Market and Financial Services) from the Commission argued that there was no single market for security services, and therefore no need for central regulation.

Nevertheless, the possibility of legislation has not been excluded. When questioned on this issue at the September 1996 European conference, the Head of Unit of DG V said that the Commission had not ruled out action in this area and it would be up to the social partners to present proposals for them to consider. Even if the Commission decided to implement common standards on some aspect of the European private security industry, the process would still take many years.

Another dimension also exists for the development of regulations that may significantly affect the UK private security industry introduced under the Social Chapter, as the Labour government ended the opt-out in 1997. There are also some developments in European law, adopted under the main treaty, that pre-date the Social Chapter and have been applicable to the UK. The Directive on working hours was introduced through the Working Time Regulations (1998) (as discussed earlier). The Directive on European Works Councils has already led Group 4, one of the largest players in the UK market, to establish its own Works Council. Another example is the 1998 *Data Protection Act*, which was introduced to implement the Data Protection Directive 1995/46/EC. It will have a significant impact on the management of information, which many private security companies routinely engage in.

However, the Directive that has had the greatest impact on the UK private security industry so far is the Acquired Rights Directive of 1977. This was implemented in the UK by the 1981 Transfer of Undertakings (Protection of Undertakings) Regulations (more commonly known as the TUPE regulations). The aim of the legislation was to protect the rights of employees in the event of a transfer of undertakings, businesses or parts of businesses. The regulations mean that the contracts of employment of individuals are automatically transferred to the new employer, as are any collective agreements. Individuals are protected from dismissal (except for economic, technical or organisational reasons) and trade union representatives must be consulted before a transfer takes place. Originally the regulations were not felt to apply to cases of contracting out, but subsequent judgements by the European Court of Justice have extended their coverage. The impact on the contract guarding sector has been dramatic as the market is essentially price driven and much business is gained by undercutting rivals through cutting labour costs. Firms that are awarded a new contract and seek to keep the same security staff are now faced with legal barriers to lowering their pay and conditions (Button and George, 1994).

European standards institutions

There are two bodies at European level concerned with the harmonisation of national standards into European Norms and the development of new European standards. CEN covers a wide range of areas including quality management, materials, mechanical engineering, building and civil engineering, healthcare, information technology, the environment and food. Sectors of the private security industry that would fall under CEN include locks, safes and security glass. The second body is CENELEC, which is concerned with electro-technical standards. CENELEC has taken a greater interest in some aspects of private security with Technical Committee (TC) 79, which has responsibility for electronic security. There are a further 12 working groups charged with various technical aspects of CCTV, alarms, access control and integrated systems (Pasco, 1996). The procedures for the adoption of European Standards are very complex and it would not be possible to do them justice in the limited space available. Those interested, are referred to Pasco (1996) and Finney (1996). What is clear however, is that in the future standards for products and services will increasingly be decided at a European level.

Other European intervention

As signatories to the European Convention on Human Rights the UK government has frequently been found in breach of the Convention by its court. In 1998 the *Human Rights Act* was passed, which, when implemented, will place the Convention directly into our law. This is almost certain to lead to a right to privacy through Common Law (Craig, 1999). This may have a significant impact on the many private security organisations engaged in the gathering of intelligence and conducting surveillance, amongst other activities.

Future challenges for policy-makers

It would seem appropriate to end this book on some of the future challenges policy-makers are likely to face. This chapter has considered how private security has been treated as a policy issue by government and other governmental organisations. Most frequently this has meant considering some form of regulation. Once the system of regulation has been implemented in England and Wales the biggest challenge will be to ensure that it is operating efficiently and effectively and where it is not, to ensure that it is reformed so that it does. There are a number of trends already identifiable that will pose challenges to future regulators and policy-makers.

The global market in security companies operating throughout the world is also likely to pose new challenges. Already there are some companies with divisions in many parts of the world — Group 4, Securitas and Securicor, for example (Johnston, forthcoming). The recent merger of Group 4 and Falck, creating one of the largest security companies in the world, may illustrate the beginning of consolidation in the market and the emergence of a few global players that dominate the security industry throughout the world. Regulating such companies at a national level may prove increasingly difficult.

In many locations communities are increasingly concerned with the fear of crime, low-level crime and disorder. Many are seeking greater uniformed presence on the streets to alleviate these fears. Already ideas and policies are emerging to address concerns that will see the extension of the role of private security and other non-police bodies in patrolling public areas (Blair, 1998; Draft Report of Policy Action Team 6, 1999; Jacobson and Saville, 1999; and Policy Action Team 6, 2000). Ensuring that these new initiatives are of the highest standard and are accountable will also be a challenge for policy-makers.

Chapter 10 illustrated the wide range of technological developments in the security products sector. It is likely that improvements in technology will continue to fuel the 'arms race' in security products, which will become ever more complex and sophisticated. Some of the trends include the increasing use of integrated systems utilising a wider range of security functions (and non-security) such as intruder alarms, CCTV, access control systems, fire alarms etc. Many of these systems will also make use of digital transmission technology, which enables images to be transmitted down telephone lines, potentially to other countries! Each of the different security technologies are also likely to become increasingly sophisticated, effective and complex to service and use, and many will replace security officers (BSIA, 1999b). Private companies may also begin to make greater use of satellite technology. Already spy satellites can be hired by private individuals to focus upon a specific area (Button, 1999). These developments will also pose new challenges for policy-makers.

These are just some of the issues that are likely to emerge and that will require a considered response by policy-makers, the industry and other stakeholders. It is hoped that this book has provided a basis for a better understanding of private security and illustrated some of the many areas that require further research because, ultimately, as the role and size of the industry expands, so it will be increasingly necessary to understand it. Without that understanding the full and effective positive contribution of private security to society is never likely to be completely realised.

References

Allen, S. (1991) *The A-Z Guide to European Manned Guarding.* London: Network Security Management.

Amnesty International (1994) *United Kingdom Cruel, Inhuman or Degrading Treatment During Forcible Deportation.* London: Amnesty International.

APEX (u.d.) *A European Charter for the Private Security Industry.* London: APEX.

Association of British Insurers (ABIn) (1994) *Guidelines Insurers' Minimum Security Requirements for Domestic Properties.* April 1994. London: ABIn.

Association of British Insurers (ABIn) (1996) *Public Affairs Regional Newsletter.* February 1996.

Association of British Investigators (ABI) (1995) *Directory.* Kingston-upon-Thames: ABI.

Association of Chief Police Officers (ACPO) (1988) *A Review of the Private Security Industry.* Unpublished Report.

Association of Chief Police Officers (ACPO) (1995a) *National Intruder Alarm Statistics 1994.* Lewes: Sussex Police.

Association of Chief Police Officers (ACPO) (1995b) *ACPO Intruder Alarms Policy 1995.* London: ACPO.

Association of Chief Police Officers (ACPO) (1995c) Memorandum of Evidence. In House of Commons Home Affairs Committee (1995) *The Private Security Industry.* Volume II. London: HMSO.

Atkins, S., Husain, S. and Storey, A. (1991) *The Influence of Street Lighting on Crime and Fear of Crime.* Crime Prevention Unit Paper 28. London: Home Office.

Audit Commission (1996) *Streetwise - Effective Police Patrol.* London: HMSO.

Baldeschwieler, J. (1993) Explosive Detection for Commercial Aircraft Security. In, Wilkinson, P. (ed.) *Technology and Terrorism.* London: Frank Cass.

Bamfield, J. (1994) Electronic Article Surveillance: Management Learning in Curbing Theft. In Gill, M. (ed.) *Crime at Work: Studies in Security and Crime Prevention.* Leicester: Perpetuity Press.

Barclay, G. C. and Tavares, C. (1999) *Information on the Criminal Justice System in England and Wales.* London: Home Office Research and Statistics Department.

Beck, A., Gill, M. and Willis, A. (1994) Violence in Retailing: Physical and Verbal Victimisation of Staff. In Gill, M. (ed.) *Crime at Work: Studies in Security and Crime Prevention.* Leicester: Perpetuity Press.

Beck, A. and Willis, A. (1994a) The Changing Face of Terrorism: Implications for the Retail Sector. In Gill, M. (ed.) *Crime at Work: Studies in Security and Crime Prevention.* Leicester: Perpetuity Press.

Beck, A. and Willis, A.(1994b) Customer and Staff Perceptions of the Role of Close Circuit Television in Retail Security. In Gill, M. (ed.) *Crime at Work: Studies in Security and Crime Prevention.* Leicester: Perpetuity Press.

Beck, A. and Willis, A. (1995) *Crime and Security - Managing the Risk to Safe Shopping.* Leicester: Perpetuity Press.

Beck, U. (1992) *Risk Society: Towards a New Modernity.* London: Sage.

Biles, D. and Vernon, J. (ed.) (1994) *Private Sector and Community Involvement in the Criminal Justice System.* Proceedings of a conference held 30 November — 2 December 1992, Wellington, New Zealand. Canberra: Australian Institute of Criminology.

Black, J. (1996) Constitutionalising Self-Regulation. *The Modern Law Review.* Vol. 59, pp 24-55.

Blair, I. (1998) *Where do the Police Fit into Policing?* Speech delivered to ACPO Conference, 16th July, 1998.

Boothroyd, J. (1988a) Private Eyes Under the Microscope. *Police Review.* 26th August, 1988.

Boothroyd, J.(1988b) Casting Off the Dirty Mac. *Police Review,* 2nd September, 1988.

Boothroyd, J.(1988c) Safe as Houses. *Police Review,* 9th September, 1988.

Borodzicz, E.P. (1996) Security and Risk: A Theoretical Approach to Managing Loss Prevention. *International Journal of Risk, Security and Crime Prevention,* Vol. 1, pp 131-143.

Bright, J. (1993) *Patrolling the Streets: A Job for the Police, the Private Security Industry or the Active Citizen.* Paper presented to the 'Policing — Private or Public' Conference at Manchester Metropolitan University, September 1993.

British Entertainment and Discotheque Association (BEDA) (1995) Memorandum of Evidence. In House of Commons Home Affairs Committee (1995) *The Private Security Industry.* Volume II. London: HMSO.

British Retail Consortium (BRC) (1998) *Retail Crime Survey 1998.* London: BRC.

British Retail Consortium (BRC)(1999) *Retail Crime Survey 1999.* London: BRC.

British Security Industry Association (BSIA) (1994a) *The UK Manned Guarding Market.* Worcester: BSIA.

British Security Industry Association (BSIA)(1994b) *Regulation of the Security Industry. A BSIA Briefing Paper-April 1994.* Worcester: BSIA.

British Security Industry Association (BSIA)(1994c) *Attitudes to the UK Manned Guarding Market.* Worcester: BSIA.

British Security Industry Association (BSIA)(1999a) *Security Direct.* Worcester: BSIA.

British Security Industry Association (BSIA)(1999b) *Manned Security Market Survey.* Worcester: BSIA.

Bryant, B. (1996) *Twyford Down - Roads, Campaigning and Environmental Law.* London: Chapman and Hall.

Buckwalter, A. (1984) *Investigative Methods.* Stoneham (USA): Butterworth.

Bunyan, T. (1976) *The History and Practice of the Political Police in Britain.* London: Quartet.

Butler, G. (1994) Shoplifters Views on Security: Lessons for Crime Prevention. In Gill, M. (ed.) *Crime at Work: Studies in Security and Crime Prevention.* Leicester: Perpetuity Press.

Button, M. (1998) Beyond the Public Gaze: The Exclusion of Private Investigators from the British Debate over Regulating Private Security. *International Journal of the Sociology of the Law,* Vol. 26, pp 1-16.

Button, M. (1999) Private Security and its Contribution to Policing: Under-researched, Under-utilised and Underestimated. *International Journal of Police Science and Management.* Vol. 2, pp 103-116.

Button, M., Brearley, N. and John, T. (1999) *New Challenges in Public Order Policing - The Professionalisation of Environmental Protest and the Emergence of the Militant Environmental Activist.* Paper Presented to the British Criminology Conference. Liverpool, 13-16 July, 1999.

Button, M. and George, B. (1994) Why Some Organisations Prefer In-house to Contract Security Staff. In Gill, M. (ed.) *Crime at Work: Studies in Security and Crime Prevention.* Leicester: Perpetuity Press.

Button, M. and George, B. (1998) Why Some Organisations Prefer Contract to In-house Security Staff. In Gill, M. (ed.) *Crime at Work: Increasing the Risk for Offenders.* Leicester: Perpetuity Press.

Button, M. and George, B. (Forthcoming) Government Regulation in the UK Private Security Industry: The Myth of Non-Regulation. *Security Journal.*

Byrne, R. (1991) *Safecracking.* London: Grafton.

Cabinet Office (1996) *The Government Response to the Deregulation Task Force Report 1996.* London: Cabinet Office.

Carratu, V. (1995) *Copyright Infringement and Product Piracy.* Paper presented to the Lisbon Security Conference, 31 May, 1995.

Civil Aviation Authority (CAA) (1976) *World Airline Accident Summary 1946-1975 Volume 1.* London: Civil Aviation Authority.

Civil Aviation Authority (CAA) (1994) *World Airline Accident Summary 1976-1993 Volume 2.* London: Civil Aviation Authority.

Clarke, M. (1992) *British External Policy Making in the 1990s.* London: Macmillan.

Clayton, T. (1967) *The Protectors.* London: Oldbourne.

Cook, P. (1995) William Spurrier and the Forgery Laws. *Holdsworth Law Review.* Vol. 17, pp 23-36.

Craig, J. D. R. (1999) *Privacy and Employment Law.* Oxford: Hart Publishing.

Critchley, T. A. (1978) *A History of Police in England and Wales.* London: Constable.

Cunningham, W. C., Strauchs, J. J. and Van Meter, C. W. (1990) *Private Security Trends 1970-2000.* Hallcrest Report II. Stoneham (USA): Butterworth Heinemann.

Cunningham, W. C. and Taylor, T. (1985) *Private Security and Police in America.* The Hallcrest Report. Portland: Chancellor Press.

Davies, S. (1996) *Big Brother - Britain's Web of Surveillance and the New Technological Order.* London: Macmillan.

de Quidt, J. (1997) The Origins and Role of the Football Licensing Authority. In, (eds) Frosdick, S. and Walley, L. (eds) (1997) *Sport and Safety Management.* Oxford: Butterworth Heinemann.

Department of National Heritage (1995) *Guide to Safety at Sports Grounds.* London: HMSO.

Department of Transport (1993) *Transport Security Division Corporate Plan 1994-95 to 1997-98.* London: Department of Transport.

Deregulation Task Force (1996) *Report 1995/96.* London: Deregulation Task Force.

Dorey, F. (1983) *Aviation Security.* London: Granada.

Draft Report of Policy Action Team 6 (1999) *Neighbourhood Wardens.* London: Home Office.

Draper, H. (1978) *Private Police.* Sussex: Harvester Press.

Drury, I. and Bridges, C. (1995) False Dawn. *Security Surveyor.* Vol. 26, No.2.

Eisner, M. A. (1993) *Regulatory Politics in Transition.* Baltimore - Maryland: John Hopkins University Press.

Ekblom, P., Simon, F. and Birdi, S. (1988) *Crime and Racial Harassment in Asian-run Small Shops: the Scope for Prevention.* Crime Prevention Unit Paper 15. London: Home Office.

Electronic Article Surveillance Manufacturers Association (EASMA) (u.d.) *Electronic Article Surveillance - A Practical Introduction for Retailers.* Beaconsfield: EASMA.

Elliot, N. (1988) *Making Prisons Work.* London: Adam Smith Institute.

Emsley, C. (1991) *The English Police.* London: Longman.

Euromonitor (1989) *The UK Security Report 1989.* London: Euromonitor.

European Commission (1996) *Social Europe, Social Dialogue - The Situation in the Community in 1995.* Luxembourg: Office for the Official Publications of the European Community.

Fieldsend, T. (1994) *Intruder Alarms the Way Forward?* London: Home Office.

Finney, J. (1996) European Standards: Publication Date Approaching. In BSIA, *Security Direct.* Worcester: BSIA.

Fischer, R. J. and Green, G. (1992) *Introduction to Security. Boston*: Butterworth-Heinemann.

Francis, J. (1993) *The Politics of Regulation.* Oxford: Blackwell.

Frosdick, S. and Walley, L. (eds) (1997) *Sport and Safety Management.* Oxford: Butterworth Heinemann.

Garland, J. and Rowe, M. (1996) Racism at Work: A Study of Professional Football. *International Journal of Risk, Security and Crime Prevention.* Vol. 1. pp 195-206.

George, B. and Button, M. (1995) Memorandum of Evidence. In House of Commons Home Affairs Committee (HAC) (1995) *The Private Security Industry.* Vol. I and II. HC 17. London: HMSO.

George, B. and Watson, T. (1992) Regulation of the Private Security Industry. *Public Money and Management.* Vol.12, pp 55-57.

Giddens, A. (1990) *The Consequences of Modernity.* Cambridge: Cambridge University Press.

Gill, K. M. (1994) Fiddling in Hotel Bars: Types, Patterns, Motivations and Prevention. In Gill, M. (ed.) *Crime at Work: Studies in Security and Crime Prevention.* Leicester: Perpetuity Press.

Gill, M. (ed.) (1994) *Crime at Work: Studies in Security and Crime Prevention*. Leicester: Perpetuity Press.

Gill, M.(1996) Risk, Security and Crime Prevention: An International Forum for Developing Theory and Practice. *International Journal of Risk, Security and Crime Prevention*. Vol. 1. pp 11-17.

Gill, M. (ed) (1998) *Crime at Work: Increasing the Risk for Offenders* Leicester: Perpetuity Press.

Gill, M. and Hart, J. (1997a) Policing as a Business: The Organisation and Structure of Private Investigation. *Policing and Society*. Vol. 7, pp 117-141.

Gill, M. and Hart, J.(1997b) Exploring Investigative Policing. *British Journal of Criminology*. Vol. 37, pp 549-567.

Gill, M., Hart, J. and Stevens, J. (1996) Private Investigators: Under-researched, Under-estimated, Under-used? *International Journal of Risk, Security and Crime Prevention*. Vol. 1, pp 305-316.

Gill, M. and Mawby, R. (1990) *Volunteers and the Criminal Justice System: A Comparative Analysis*. Milton Keynes: Open University Press.

Gilling, D. (1997) *Crime Prevention - Theory Policy and Politics*. London: UCL Press.

Greenouff, A. (u.d.) *The Role of the Vehicle Security Installation Board*. Unpublished Paper.

Group 4 (1992) *Training Services Manual*.

Handelman, S. (1994) *Comrade Criminal*. London: Michael Joseph.

Handford, M. (1994) Electronic Tagging in Action: A Case Study in Retailing. In Gill, M. (ed) *Crime at Work: Studies in Security and Crime Prevention*. Leicester: Perpetuity Press.

Harding, R. W. (1997) *Private Prisons and Public Accountability*. Buckingham: Open University Press.

Health and Safety Commission, Home Office and Scottish Office (1993) *Guide to Health, Safety and Welfare at Pop Concerts and Similar Events*. London: HMSO.

Hearnden, K. (1993) *The Management of Security in the UK*. Loughborough: Centre for Extension Studies, University of Loughborough and SITO.

Hearnden, K. (1995) Multi-tasking in British Business: A Comparative Study of Security and Safety Managers. *Security Journal*. Vol. 6. pp 123-132.

Heims, P. (1982) *Countering Industrial Espionage*. Leatherhead: 20th Century Security Education.

Hibberd, M. and Shapland, J. (1993) *Violent Crime in Small Shops.* London: Police Foundation.

HM Inspectorate of Prisons (1995) *Report of an Unannounced Short Inspection By HM Inspectorate of Prisons - Immigration Detention Centre Campsfield House.* London: Home Office.

HM Prison Service (1995) *Corporate Plan 1995-98.* London: HM Prison Service.

Home Office (u.d.) Document on Training and Registration Schemes. Appendix 3 In Bedborough, P. *Public Order and Social Control on Licensed Premises - An Analysis of the Changing Role of Door Supervisors.* Unpublished Dissertation at Nottingham Trent University.

Home Office (1979) *The Private Security Industry: A Discussion Paper.* London: HMSO.

Home Office (1990a) *Working Group on Regulation of Safe Deposit Centres.* Unpublished Home Office Document.

Home Office (1990b) *Court Escorts, Custody and Security: A Discussion Paper.* London: Home Office.

Home Office (1991) *Private Security Industry Background Paper.* Unpublished Paper.

Home Office (1993a) *Wheel Clamping on Private Land.* London: HMSO.

Home Office (1993b) *Disclosure of Criminal Records for Employment Vetting Purposes - A Consultation Paper by the Home Office.* London: HMSO.

Home Office (1994a) *Report of the Gaming Board for Great Britain 1993/94.* London: HMSO.

Home Office (1994b) *CCTV - Looking Out for You.* London: HMSO.

Home Office (1994c) *Part II: Police, Drug Misusers and the Community.* London: HMSO.

Home Office (1995a) Memoranda of Evidence Submitted to the Home Affairs Committee Inquiry into the Private Security Industry. In House of Commons Home Affairs Committee. *The Private Security Industry.* Volume II. London: HMSO.

Home Office (1995b) *Prison Statistics England and Wales 1993.* Cm 2893. London: HMSO.

Home Office (1996a) *Regulation of the Contract Guarding Sector of the Private Security Industry.* London: Home Office.

Home Office (1996b) *On the Record.* Cm 3308. London: HMSO.

Home Office (1997) *Criminal Statistics England and Wales 1996.* Cm 3764. London: The Stationery Office.

Home Office (1998) *Fairer, Faster and Firmer - A Modern Approach to Immigration and Asylum.* Cm 4018. London: The Stationery Office.

Home Office (1999) *The Government's Proposals for the Regulation of the Private Security Industry in England and Wales.* Cm 4254. London: The Stationery Office.

Hou, C. and Sheu, C. J. (1994) A Study of the Determinants of Participation in a Private Security System Among Taiwanese Enterprises. *Police Studies.* Vol. 17, pp 13-23.

Houghton, S. (1994) The Why and What and How of Access Control. *International Fire and Security Product News.* Vol. 19. No 5.

House of Commons Defence Committee (HCDC) (1984) Second Report. *The Physical Security of Military Installations in the United Kingdom.* HC 397-I. London: HMSO.

House of Commons Defence Committee (HCDC)(1990) Sixth Report. *The Physical Security of Military Installations in the United Kingdom.* HC 171. London: HMSO.

House of Commons Defence Committee (HCDC)(1996) Eighth Report. *Ministry of Defence Police and Guarding.* HC189. London: HMSO.

House of Commons Home Affairs Committee (HAC) (1995) *The Private Security Industry.* Vol. I and II. HC 17. London: HMSO.

Ignatieff, M. (1978) *A Just Measure of Pain.* London: MacMillan.

Immigration and Nationality Department (1994) *Annual Report 1994.* London: Home Office.

Inspectorate of the Security Industry (ISI) (1998) *Register of Manned Security Companies.* Issue 5 - January 1998. Droitwich: ISI.

Institute of Professional Investigators (IPI) (1992) *Survey of the UK Investigatory Industry and Profession.* Blackburn: IPI.

Jacobson, J. and Saville, E. (1999) *Neighbourhood Warden Schemes: An Overview.* Crime Reduction Research Series Paper 2. London: Home Office.

Jacques, C. (1994) Ram Raiding: The History, Incidence and Scope for Prevention. In Gill, M. (ed) *Crime at Work: Studies in Security and Crime Prevention.* Leicester: Perpetuity Press.

James, A. L., Bottomley, A., K., Liebling, A. and Clare, E. (1997) *Privatising Prisons - Rhetoric and Reality.* London: Sage.

Jenkins, B. M. (1999) Aircraft Sabotage. In, Wilkinson, P. and Jenkins, B., M. (eds) *Aviation Terrorism and Security.* London: Frank Cass.

Jewell, H. M. (1972) *English Local Administration in the Middle Ages.* Newton Abbot: David Charles.

Johnston, L. (1991) Privatisation and the Police Function: From 'New Police' to 'New Policing'. In Reiner, R. and Cross, M. (eds) *Beyond Law and Order: Criminal Justice Policy and Politics into the 1990's.* Basingstoke: MacMillan.

Johnston, L. (1992a) *The Rebirth of Private Policing.* London: Routledge.

Johnston, L. (1992b) An Unseen Force: The MoD Police in the UK. *Policing and Society.* Vol. 3, pp 23-40.

Johnston, L. (1994) Privatisation: Threat or Opportunity. *Policing.* Vol. 10, pp 14-22.

Johnston, L. (2000) *Policing Britain: Risk, Security and Governance.* London: Longman.

Johnston, L. (Forthcoming) *Transnational Policing.*

Joint Council for the Welfare of Immigrants (JCWI) (1995) *Memorandum to the Home Affairs Committee Inquiry into the Private Security Industry.* Unpublished Submission.

Jones, P. H. (1997) *Retail Loss Control.* Oxford: Butterworth Heinemann.

Jones, T. and Newburn, T. (1994) *How Big Is the Private Security Industry?* London: Policy Studies Institute.

Jones, T. and Newburn, T. (1998) *Private Security and Public Policing.* Oxford: Clarendon Press.

Jordans and Son (1987) *Britain's Security Industry.* London: Jordans and Son.

Jordans and Son (1992) *Britain's Security Industry.* London: Jordans and Son.

Joselyn, J. (1994) Private Investigations - Inside Story. *Security Industry.* December 1994.

Kakalik, J. and Wildhorn, S. (1971a) *Private Police in the United States, Findings and Recommendations.* Vol. 1. Washington DC: Government Printing Office.

Kakalik, J. and Wildhorn, S.(1971b) *The Private Police Industry: Its Nature and Extent.* Vol. 2, Washington DC: Government Printing Office.

Kakalik, J. and Wildhorn, S.(1971c) *Current Regulation of Private Police: Regulatory Agency Experience and Views.* Vol. 3, Washington DC: Government Printing Office.

Kakalik, J. and Wildhorn, S.(1971d) *The Law and Private Police.* Vol. 4, Washington DC: Government Printing Office.

Kakalik, J. and Wildhorn, S.1971e) *Special Purpose Public Police.* Vol. 5, Washington DC: Government Printing Office.

Keynote (1993) *Market Review UK Security.* Hampton: Keynote.

Keynote (1996) *DTI Competitive Analysis of the Lock Industry in the UK.* London: Keynote.

Latter, R. (1992) *The Terrorist Threat to Business.* Wilton Park Paper 62. London: HMSO.

Liardet, V. (1995) Mechanical Protection. *Security Surveyor* Vol. 26, No 1.

Lilly, J. R. and Knepper, P. (1992) An International Perspective on the Privatisation of Corrections. *The Howard Journal of Criminal Justice.* Vol. 31, pp 174-191.

Linebaugh, P. (1991) *The London Hanged - Crime and Civil Society in the Eighteenth Century.* London: Penguin.

Litherland, J. (1994) Air Cargo Security. *Intersec.* Vol. 4, No. 10 (Aviation Security Supplement).

Loader, I. (1997) Thinking Normatively About Private Security. *Journal of Law and Society.* Vol. 24, pp 377-94.

Local Government Information Unit (LGIU) (1996) *Candid Cameras A Report on Closed Circuit Television.* London: Local Government Information Unit.

Loss Prevention Certification Board (1999) *Approved Fire and Security Products and Services 1999.* London: Loss Prevention Certification Board.

Loveday, B. (1992) Right Agendas: Law and Order in England and Wales. *International Journal of the Sociology of Law.* Vol. 20, pp 297-319.

Loveday, B. (1994) Government Strategies for Community Crime Prevention Programmes in England and Wales: A Study in Failure. *International Journal of the Sociology of Law.* Vol. 22, pp 181-202.

Loveday, B. (1995) Contemporary Challenges to Police Management in England and Wales: Developing Strategies for Effective Service Delivery. *Policing and Society.* Vol. 5, pp 281-302.

McGuire, F. G. (1989) *Aviation Security Strategies for the '90s.* Washington: McGraw-Hill.

McManus, M. (1995) *From Fate to Choice: Private Bobbies, Public Beats.* Aldershot: Avebury.

Maguire, M. (1994) Crime Statistics, Patterns, and Trends: Changing Perceptions and their Implication. In Maguire, M., Morgan, R. and Reiner, R. (eds) *The Oxford Handbook of Criminology.* Oxford: Oxford University Press.

Mair, G. and Mortimer, E. (1996) *Curfew Orders and Electronic Monitoring.* Home Office Research Study 163. London: Home Office.

Mair, G. and Nee, C. with Barclay, G. and Wickham, K. (1990) *Electronic Monitoring: The Trials and their Results.* Home Office Research Study 120. London: HMSO.

Manto, S. E. (1993) Airline Security: Will the Public Pay? Results of a Limited Survey among Chicago-Area Air Travellers. *Security Journal*. Vol. 4, pp 221-234.

Manunta, G. (1996) The Case Against: Security Management is Not a Profession. *International Journal of Risk, Security and Crime Prevention*. Vol. 1, pp 233-240.
Manunta, G. (1999) What is Security? *Security Journal*. Vol. 12, pp 57-66.

Mars, G. (1982) *Cheats at Work: An Anthropology of Workplace Crime*. London: Allen and Unwin.

Master Locksmiths Association (1994) *Constitution and Byelws Edition One*. Daventry: MLA.

Mayhew, P., Maung, N. A. and Mirrlees-Black, C. (1993) *The 1992 British Crime Survey*. London: HMSO.

MBD (1994) *The UK Security Equipment Market Development*. Manchester: Market and Business Development Ltd.

Melville-Lee, W. L. (1901) *A History of Police in England*. London: Metheun and Co.

Merari, A. (1999) Attacks on Civil Aviation: Trends and Lessons. In Wilkinson, P. and Jenkins, B. M. (eds) *Aviation Terrorism and Security*. London: Frank Cass.

Mirrlees-Black, C., Mayhew, P. and Percy, A. (1996) *The 1996 British Crime Survey England and Wales*. London: Government Statistical Service.

Monk, E. (1987) *Keys. Their History and Collection*. Ayesbury: Shire Publications.

Moore, K. C. (1991) *Airport, Aircraft and Airline Security*. Boston: Butterworth-Heinemann.

Moore, R. H. (1995) Private Security in the Twenty First Century: An Opinion. *Journal of Security Administration*. Vol. 18, pp 3-20.

Moran, P. and Alexandrou, A. (1994) *The Police Force of the Future A Security Risk?* Paper presented to the conference. Changing Perceptions of Risk: Implications for Management. Bolton. 27 February - 1 March, 1994.

Morris, T. (1989) *Crime and Criminal Justice Since 1945*. London: Basil Blackwell.

Moyle, P. (1993) Privatisation of Prisons in New South Wales and Queensland: A Review of Some Key Developments in Australia. *The Howard Journal of Criminal Justice*. Vol. 32, pp 231-50.

Moyle, P. (ed) (1994) *Private Prisons and Police - Recent Australian Trends*. Leichhardt (Australia): Pluto Press.

Murphy, D. J. I. (1986) *Customers and Thieves*. Aldershot: Gower.

Murray, C. (1996) The Case Against Regulation. *International Journal of Risk, Security and Crime Prevention*. Vol. 1, pp 59-62.

National Advisory Committee on Criminal Standards and Goals (1976) *Private Security. Report of the Task Force on Private Security*. Washington DC: Government Printing Office.

National Approval Council for Security Systems (NACOSS) (1993) *Official List of NACOSS Recognised Firms*. Maidenhead: NACOSS.

National Approval Council for Security Systems (1998) *Official List of NACOSS Registered Firms*. Twelfth Edition. Maidenhead: NACOSS.

National Audit Office (1994) *Wolds Remand Prison*. London: HMSO.

National Audit Office (1995) *Entry into the United Kingdom*. HC 204 Session 1994-95. London: HMSO.

National Fencing Training Authority (NFTA) (1995) *Fencing Industry Labour Market Analysis*. Burton-upon-Trent: NFTA.

Nedin, P. (1995) Special Limousine Protection. *Intersec*. Vol. 5, No. 3.

Nottingham Safer Cities Project (1990) *Community Safety in Nottingham City Centre - Report of the Steering Group*.

Owen, W. (1995) Bodyguards: Who is Making Money Out of the Fantasy? *Intersec*. Vol. 5, No. 11/12.

Oxley, J. C. (1993) Non-Traditional Explosive: Potential Detection Problems. In Wilkinson, P. (ed.) *Technology and Terrorism*. London: Frank Cass.

Palmer, S. G. (1988) *Police and Protest in England and Wales 1789-1850*. Cambridge: Cambridge University Press.

Pasco, I. K. (1996) Standards in the New Europe. In BSIA, *Security Direct 96/97*. Worcester: BSIA.

Pease, K. (1994a) Crime Prevention. In Maguire, M., Morgan, R. and Reiner, R. (eds) *The Oxford Handbook of Criminology*. Oxford: Oxford University Press.

Pease, K. (1994b) Forget the Ideology: What Could Privatisation Contribute to Prison Reform? In Martin, C. (ed) *Contracts to Punish: Private or Public?* Report of a conference organised by the Institute for the Study and Treatment of Delinquency held in Manchester on 24 November, 1994.

Phillips, D. (1973) *Skyjack The Story of Air Piracy*. London: Harrup.

Phillips, S. and Cochrane, R. (1988) *Crime and Nuisance in the Shopping Centre: A Case Study in Crime Prevention*. Crime Prevention Unit Paper 16. London: Home Office.

Pitkin, H. (1984) *Fortune is a Woman.* Berkeley: University of California Press.

Police Foundation and Policy Studies Institute Independent Inquiry (PF/PSI) (1994) *Independent Committee of Inquiry into the Role and the Responsibilities of the Police, Discussion Document.* London: PF.

Police Foundation and Policy Studies Institute Independent Inquiry (PF/PSI) (1996) *The Role and Responsibilities of the Police.* London: PF/PSI.

Police Scientific and Development Branch (PSDP) (1995) *Annual Report 1994/95.* St Albans: PSDB.

Policy Action Team 6 (2000) *Neighbourhood Wardens.* London: Home Office.

Poole, R. (1991) *Safer Shopping.* Volume 1. Birmingham: West Midlands Police.

Prenzler, T. and Sarre, R. (1999) A Survey of Security Legislation and Regulation Strategies in Australia. *Security Journal* Vol. 12, pp 7-17.

Priestly, H. (1968) *Voice of Protest.* London: Leslie Frewin.

Public Accounts Committee (PAC) (1995) *Ministry of Defence: Fraud in Defence Procurement.* London: HMSO.

Radzinowicz, L. (1956) *A History of the English Criminal Law.* Vol. II. London: Stevens.

Ramsey, M. and Newton, R. (1991) *The Effect of Better Street Lighting on Crime and Fear: A Review.* Crime Prevention Unit Paper 29. London: Home Office.

Reese, P. (1986) *Our Sergeant - The Story of the Corps of Commissionaires.* London: Leo Cooper Ltd.

Reiner, R. (2000) Police Research. In. King, R. D. and Wincup, E. (eds) *Doing Research on Crime and Justice.* Oxford: Oxford University Press.

Reynolds, E. A. (1998) *Before the Bobbies - The Night Watch and Police Reform in Metropolitan England 1720-1830.* Stanford: Stanford University Press.

Ruddock, J. (1994) *Privatising Prisons: A Dangerous Diversion.* London: Labour Party.

Ryan, M. and Ward, T. (1989) *Privatization and the Penal System.* Milton Keynes: Open University Press.

Sarre, R. (1994) The Legal Powers of Private Police and Security Providers. In Moyle, P. (ed.) *Private Prisons and Police - Recent Australian Trends.* Leichhardt, NSW: Pluto Press.

Schaffer, M. B. (1999) The Missile Threat to Civil Aviation. In Wilkinson, P. and Jenkins, B. M. (eds) *Aviation Terrorism and Security.* London: Frank Cass.

Schiffi, P. (1989) *The Guard Service Industries in the FRG-with Changing Requirements.* Paper given at ASIS conference, Frankfurt, 31 May - 2 June, 1989.

Security Industry Training Organisation (SITO) (1993a) *The Training Directory.* Worcester: SITO.

Security Industry Training Organisation (SITO)(1993b) *A Review of the Electronic Security Industry.* Worcester: Young Samuel and Chambers Ltd.

Security Management Today (SMT) (1996) *International Fire and Security Directory.* London: SMT.

Shearing C. D. (1981) Private Security in Canada: Some Questions and Answers. In. McGrath, W. T. and Mitchell, M. P. (eds) *The Police Function in Canada.* Toronto: Methuen.

Shearing, C. D. and Stenning, P. C. (1981) Modern Private Security: Its Growth and Implications. In. Tonry, M. and Morris, N. (eds) *Crime and Justice An Annual Review of Research. Volume 3.* Chicago: University of Chicago Press.

Shearing, C. D. and Stenning, P. C. (1982) *Private Security and Private Justice.* Montreal: The Institute for the Research on Public Policy.

Shearing, C. D. and Stenning, P. C. (eds) (1987) *Private Policing.* Thousand Oaks (California): Sage.

Shearing, C. D., Stenning, P. C. and Addario, S. M. (1985) Police Perceptions of Private Security. *Canadian Police College Journal.* Vol. 9, pp 127-153.

Shichor, D. (1995) *Punishment for Profit.* Thousand Oaks (California): Sage.

Shortt, J. (1995) Jim Shortt's Bodyguard Course Part I. *Combat and Militaria,* March 1995.

Simonsen, C. E. (1996) The Case For: Security Management is a Profession. *International Journal of Risk, Security and Crime Prevention.* Vol. 1, pp 229-232.

Slaughter, P. G. (1995) *VIP Protection and Personal Security.* Paper presented to the Lisbon Security Conference, 29 May, 1995.

Smith, J. P. (1996) Explaining Crime Trends. In. Saulsbury, W., Mott, J. and Newburn, T. (eds) *Themes in Contemporary Policing.* London: Policy Studies Institute.

Somerson, I. S. (1996) *Risk Assessment and Security Program Design.* Paper Presented to ASIS Conference, Atlanta, 11 September, 1996.

South, N. (1988) *Policing for Profit.* London: Sage.

South, N. (1989) Reconstructing Policing: Differentiation and Contradiction in Post-War Private and Public Policing. In Mathews, R. (ed) *Privatising Criminal Justice.* London: Sage.

Steer, A. (1995) Qualify, Verify and Regulate. In. BSIA, *Security Direct 95*. Worcester: BSIA.

Stenning, P. C. and Shearing, C. D. (1981) The Quiet Revolution: The Nature, Development and General Legal Implications of Private Security in Canada. *Criminal Law Quarterly*. Vol. 22, pp 220-248.

Storch, R. (1975) The Plague of Blue Locusts: Police Reform and Popular Resistance in Northern England 1840-57. *International Review of Social History*. Vol. 20, pp 61-90. Swann, D. (1989) The Regulatory Scene: An Overview. In Button, K. and Swann, D. (eds) *The Age of Regulatory Reform*. Oxford: Clarendon Press.

Ta'eed, O. (1995) CCTV - Then and Now a Lesson in Change Management. *CCTV Today*. Vol. 2, No. 6.

Taylor, D. (1997) *The New Police in Nineteenth Century England - Crime, Conflict and Control*. Manchester: Manchester University Press.

Trofymowych, D. (1993) *Private Policing in Canada A Review*. Unpublished Paper.

Underwood, S. (1993) *The People Business*. Oxford: CPL Books.

Union of Shop, Distributive and Allied Workers (USDAW) (1995) *Parliamentary Briefing - Criminal Injuries Compensation*. Manchester: USDAW.

UNISON (u.d.) *Security Services - UNISON CCT Sector Analysis*. London: UNISON.

Upton, M. (1995) *Sports, Show and Stadia Security*. Paper Presented to the Lisbon Security Conference, 31 May, 1995.

US Department of Transport (1993) *Audit of Airport Security Federal Aviation Administration*. Unpublished Report.

Vidal, J. (1997) *McLibel - Burger Culture on Trial*. London: MacMillan.

Wall, D. (1998) *The Chief Constables of England and Wales*. Aldershot: Ashgate.

Wallis, R. (1993a) *Combating Air Terrorism*. Washington: Brasseys.

Wallis, R. (1993b) Aviation Security. In Wilkinson, P. (ed) *Technology and Terrorism*. London: Frank Cass.

Wallis, R. (1999) The Role of International Aviation Organisations in Enhancing Security. In. Wilkinson, P. and Jenkins, B. M. (eds) *Aviation Terrorism and Security*. London: Frank Cass.

Webb, B. and Webb, S. (1922) *English Prisons Under Local Government*. London: Longman, Greens and Co.

Weber, M. (1964) *The Theory of Social and Economic Organisation.* New York: Free Press.

Welch, C. (1995) *Teenage Wasteland - The Early Who.* Reading: ECG.

Wilkinson, P. (1993) Designing an Effective International Security System. In Wilkinson, P. (ed) *Technology and Terrorism.* London: Frank Cass.

Wilkinson, P. (1996) Report. *In. Rt Hon Lord Lloyd of Berwick, Inquiry into Legislation Against Terrorism.* Vol. II. Cm 3420. London: The Stationery Office.

Wilkinson, P. (1999) Enhancing Global Aviation Security. In Wilkinson, P. and Jenkins, B. M. (eds) *Aviation Terrorism and Security.* London: Frank Cass.

Williams, J. (1985) In Search of the Hooligan Solution. *Social Studies Review.* Vol. 1, pp 3-5.

Williams, D., George, B. and MacLennan, E. (1984) *Guarding Against Low Pay - The Case for Regulation in Contract Security.* London: Low Pay Unit.

Young, P. (1987) *The Prison Cell: The Start of a Better Approach to Prison Management.* London: Adam Smith Institute.

Younger, K. (1972) *Report on the Committee on Privacy.* Cm 5012. London: HMSO.

INDEX